T0196316

LOST IN WONDER

LOST in wonder

IMAGINING SCIENCE
and OTHER MYSTERIES

Colette Brooks

COUNTERPOINT · BERKELEY

Copyright © 2010 by Colette Brooks.
All rights reserved under International and
Pan-American Copyright Conventions.

All images are used courtesy of the author
unless otherwise noted.

Library of Congress Cataloging-in-Publication Data

Brooks, Colette.
Lost in wonder : imagining science and other mysteries
/ by Colette Brooks.
p. cm.
ISBN 978-1-58243-572-5
1. Technology—History. 2. Science—History.
3. Research—History. I. Title.
T15.B6828 2010
509—dc22
2010003257

Book design by Gopa & Ted2, Inc.
Printed in the United States of America

COUNTERPOINT
2560 Ninth Street
Suite 318
Berkeley, CA 94710

www.counterpointpress.com

For Trav
our own supernova

LOST IN WONDER

IMAGINE THE FUTURE: One might picture workers in crisp white lab coats, next to bubbling glass beakers, monitoring gauges and needles as they dart around dials. There may be a rack of pipettes or an autoclave in the room. The attendants are purposeful, making progress in some important but obscure endeavor. Chances are it will change our lives, whatever it is.

In this world, all is pristine. But the future wasn't always antiseptic.

In an earlier time, it would have seemed dark, coming to pass in underground caverns with steep, narrow stairs cut from damp walls. One would descend into the dankness, the way lit by sparks flying from a tangle of metal and wires. A solitary figure would be minding the mechanism; each spectacular eruption would send his misshapen assistant scurrying across the floor. Visitors would be cautioned not to speak of what they'd seen to anyone, an unnecessary prohibition since others would hardly have believed it.

Each of these visions revolves around secret practices

and rites known only to initiates. Cultures have always had special terms for those who worked at the far edge of knowledge—mystics, shamans, alchemists, magicians, priests.

Now we call them scientists, and they still scare us.

Two types of people seem to have emerged in the modern world, each with starkly dissimilar interests and acuities. Those on one side prefer words and pictures, while those on the other are comfortable only with numbers and data. An anecdote to one is a *non-replicable observation* to another.

There it is: two cultures, drifting apart, with the gap ever widening.

It's a crisis of sorts but it's not new; its roots lie in the seventeenth century, when a few inquisitive people announced that the book of nature had been written in the language of mathematics, though few could hope to read it. The incurious or the unqualified would be left behind, blind to the beauty and power of the emerging worldview.

The new outlook emboldened many.

Who would have thought to quantify the wind

before Sir Francis Beaufort did so in 1805? His wind scale proposed a new way of thinking about natural phenomena, its numerical rankings promising a new kind of precision.

At 2, a light breeze, the wind can be felt on one's face. At 6, a strong breeze, large branches move, telegraph wires whistle, and umbrellas are hard to hold. Beyond that, at 15, or 16, one can expect that nothing will remain upright.

It's an appealing approach; adjectives like *fierce* and *gusty* have been overworked, after all, and might be ready to be retired. Numbers offer a kind of clarity, even a kind of power.

But they aren't for the fainthearted.

One might build on Beaufort's work, design an early-warning system of sorts that measures attitudes towards science and technology, tracking levels of distress as they rise among ordinary people caught, ill-equipped, in the heart of the storm.

At 2, short newspaper articles about science appear intermittently, announcing developments that spur mild curiosity and casual mention in conversation. At 6, debates about global warming erupt that nonexperts cannot evaluate; worrisome reports surface of DNA predictors of disease; and planets one cherished as a child lose their place in the solar system. At 15, textbooks are outdated before they can appear in print; advances occur in waves, and the average person, increasingly queasy, finds the world transformed by forces he or she cannot understand, until it is utterly unfamiliar.

At that point, who would still be standing?

Maybe there's a way to bridge the gap between experts and everybody else. One might tell stories about numbers, or at the very least study available specimens at close range, examine the historical record, perform "thought experiments," practice some paradigm shifts. Visual aids might help.

Those on both sides of the divide will scoff at one's lack of credentials. And that's true, it's too late to get another degree, or solve the twin primes problem. Certainly the Riemann hypothesis is now out of reach. One's education goes only so far.

At this late date, maybe the most well-intentioned humanist can only hope to become a "Sunday scientist" of sorts.

Still, someone's got to make the effort.

1

Two FIGURES STAND at a blackboard: One's hand is smudged with chalk, he's working the piece down to its nub. The younger man is observing, taut and attentive. It may be some years before he holds the chalk himself. Those still struggling with simple concepts (numerator, denominator) might only recognize the

equal sign in the middle of each equation, or the parentheses that hold letters and numbers in their grip. Others may say the scribbles are like marks on a canvas, a modern painting. And surely science has something in common with art. But metaphor isn't rigorous enough, it thwarts a direct relation to the data. It's science as a second language—a halting translation that can't help one grasp the subtle expressivity of the equations.

Maybe a doctorate is advisable after all.

But keep looking. The board isn't really black, it's more a washed-out white. Now the chalk traces begin to speak for themselves. Someone has been stumbling, erasing whole swaths of symbols, writing and then erasing again. And that's familiar—indecision, second-guessing, a faint suggestion of remorse.

It's encouraging that even experts are sometimes unsure of themselves.

More points of connection can be found in even the most forbidding scientific biography. Sir Isaac Newton, greatest mind in a thousand years, made his own toys as a child and put mice to work in his miniature flour mills. As a young man, he played with prisms and confessed himself inordinately drawn to their vivid colors;

he had to shake off such enticements in order to proceed to the more sober study of optics. Later, in a melancholy mood, he admitted to making himself ill when thinking about the motion of the moon. And he was vexed by critics who demanded more from his work than the data would support, *as if it were a crime,* he observed, *to let Uncertainties alone.*

So he was sensitive, moody, probably didn't get along well with others.

Still, he was said to have laughed, once in his life, when an aspiring reader of his *Philosophiae naturalis principia mathematica* wondered if he ought to master Euclid's *Principles of Geometry* first. As if one could

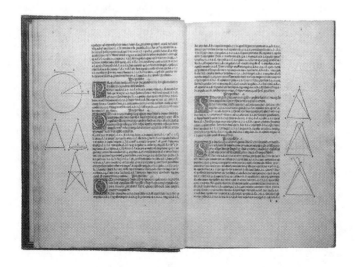

skip the great foundational text and move straight to the latest thinking, as if anyone interested in shortcuts could hope to understand either. No wonder Newton laughed.

Of course, it takes a scientific cast of mind just to understand the joke.

In 1942 Oliver La Farge famously declared that *scientists are lonely men*. He may have stumbled upon something.

Consider the case of Sir Henry Cavendish, a morbid solitary who hid in his seventeenth-century laboratory and was said to have spoken as little as possible to his own servants. Among his many accomplishments, he weighed the Earth and determined the correct composition of water, overturning the beliefs of two thousand years, but he lived a life so *barren of incident* that even his 1851 biographer was taken aback: *A more eventless life, according to the ordinary judgment of mankind, could scarcely be conceived*. Little wonder, then, that Sir Henry's own memoir was entitled *Experiments on Air*, and read before the Royal Society. Perhaps he suspected that fellow scientists might be the only ones who would care.

Three hundred years after Sir Henry, children lucky

enough to get a Lionel Chemlab for Christmas will perform experiments on the chemistry of water by themselves, determining in a few minutes what it took Cavendish a lifetime to realize. They will have to be reminded that real science isn't easy. Institutions as imperious as the Smithsonian will display cautionary placards alongside antique chemistry sets: DISCOVERIES SOMETIMES APPEAR ONLY AFTER WEEKS OR MONTHS OR YEARS OF REASONING, GUESSING, EXPERIMENTING, TRIAL AND ERROR, ACCIDENTS, TINKERING, AND BLIND LUCK. SCIENTIFIC DISCOVERY IS NOT AN INSTANT REVELATION.

But if they persevere, if they can cope with the tedium and tolerate the uncertainties, young tinkerers may find themselves in thrall to the mysteries of the natural world. They might devote their entire lives to such work, and if they are in the end shut off, without friends or even family, the trade-off might be worth it.

SOME OF THE CHILDREN WHO PLAYED WITH TOYS LIKE THESE IN THE 1930S WENT ON TO OTHER SCIENCE PROJECTS LATER IN LIFE, SUCH AS CRACKING THE GENETIC CODE AND UNLOCKING THE SECRETS OF THE ATOM.

Accidents and blind luck: Science is studded with the salutary effects of chance, great feats achieved almost

inadvertently. One might think that even the hobbyist, if alert, has a chance to get lucky. A Dutch spectacle-maker in the seventeenth century, playing around, puts two of his lenses together quite by accident, and Galileo Galilei, in Padua, hears word of *a glass by means of which distant objects might be seen as distinctly as if they were nearby.* He sets about to build his own instrument and points it at the moon, the glassy sphere that philosophers consider a heavenly example of perfection. In this new light Galileo sees that the moon isn't smooth, but pocked and cratered, like the Earth itself.

He is entranced; he has seen something that no one else has ever looked upon. He takes his glass out night after night, in a fervor, even when the object of his gaze withholds itself. *I waited for the next night with the most intense longing, but I was disappointed of my hope, for the sky was covered with clouds.*

But there are skeptics. Some simply refuse to look through the glass, others suspect sleight of hand, a deceit of some kind; perhaps the scientist, like a sorcerer, has placed the images inside the instrument itself. Galileo writes in frustration to a friend, one of the few in the world who can understand him. *I wish, my dear Kepler, that we could have a good laugh together at the extraordinary stupidity of the mob.* He marvels at the obstinacy of those who have devoted their lives to the desiccated ideas of Aristotle, afraid to look at the very world they live in. He wonders: *Why must I wait so long before I laugh with you?*

He makes more of the leather tubes we now call *telescopes.* Some will be preserved, for posterity, looking childishly simple to a more sophisticated age.

In private, he admits to a depth of emotion about his work that his worldliness belies. *I am,* he says, *quite beside myself with wonder.*

In a later century, as men from Earth first walk on the surface of its rocky satellite, descendants of those early naysayers will insist that they too have been tricked, this time by the government. And the ghost of Galileo will have another laugh.

White shirts and pocket protectors, skinny ties and buzz cuts, laminated identity cards looped around their necks: It's a roomful of rocket scientists, the very avatars of the hyperelevated IQ. They seem to be celebrating a successful launch at some point in the 1960s. A fellow in the background looks almost exactly like the figure at the blackboard in the earlier photo. But

that's unlikely, it can't be the same man moving from lab to lab, field to field, offering his services almost selflessly to science.

Still, the coincidence is uncanny.

There's Wernher von Braun, binoculars around his neck, slapping a colleague's back. His obsession with space began in the 1920s, on a small island in the North Sea, where he read science fiction as a boy and wrote stories about rockets powerful enough to leave

Earth's orbit. His mother, a baroness, gave her son a telescope so that he could study the nighttime sky. At sixteen, he launched a toy wagon powered by firecrackers. As an adult he would design the Saturn V, the rocket as long as a football field that propelled the Apollo flights to the moon.

Lindbergh didn't fly the Atlantic to get to Paris, von Braun once observed. He knew, at root, that science is not a practical pursuit; it has more to do with dreams.

Is there such a thing as a born scientist? Someone whose very temperament sets one apart?

One might look to Robert Boyle, the fourteenth child of affluent Elizabethans. With money, and a swarm of siblings engaged in more conventional pursuits, he was free to indulge his fascination for *chymistry*. It wasn't what most people in the seventeenth century would have done with wealth. Indeed, many of his contemporaries observed that the *talle and slender* figure was strange. He was clearly engaged in the darker arts, with fires burning night and day in his home. He was impervious to comfort—*His very Bed-Chamber was so crowded with Boxes, Glasses, Potts, Chemical & Mathematical Instruments; Books and Papers that there was but roome for a few Chaires.* He consulted a thermometer before deciding what to wear—why trust vague intuitions about weather when one could establish a direct correlation between temperature and clothing? And he described himself as skeptical, with proud defiance.

He prized his experiments above all earthly things, a judgment damning enough to end up a cautionary note in his funeral service. Of course, his contemporaries couldn't understand why he felt that a lifetime wasn't long enough for his work.

There was so much to be done. It was up to him to introduce experimental technique to a world still shrouded in superstition. Someone had to challenge the archaic conception of matter as earth, water, fire, and air; it was a hopelessly crude categorization, it didn't explain anything, at least nothing he wanted to know.

Someone had to invent the air pump, an artificial vacuum. Sometimes he sacrificed living creatures to his studies, suffocating a bird or a mouse in the airless chamber as he watched it expire alongside a flickering candle. The natural world was awash in cruelty; why should science deprive itself of such an able instrument?

Someone had to acknowledge that true science required a tolerance, if not a talent, for self-recrimination. His successes were easy enough to chronicle, but he kept accounts of his experimental missteps and dead ends as well. *An ineffectual attempt to kindle Sulpher in our Vacuum; An unprosperous attempt to make Flame kindle Camphire without the help of Air.* It was necessary to admit error, no honest practitioner can *pretend to knowledge.*

One might say he turned disappointment into a defining stage of scientific development.

In "New Experiments About Explosions" he records that he blows things up, and blows them up again if the proper protocols haven't been observed: *my face being accidentally turn'd to remove a Light that I feared might disturb us, I could not see the flash my self, and therefore caus'd the Experiment to be made once more, to ground my narrative upon my own observation.* It was something new, a kind of ethic of experimentation; no surprise, then, that he was an admirer of *the great stargazer Galileo.*

In "The History of Cold" he establishes that ice evaporates, once melted, explaining one more mystery in the host of natural processes no one had divined before.

Some experiments are devised simply to entertain.

He puts a clear solution in a glass vessel and adds drops of one substance to change it to deep orange, then adds something else to return the liquid to its original state. It's such a simple procedure that it can hardly be considered a trick. But those with no grounding in the new chemistry are confounded, precursors of the slack-jawed spectators who will gape at new developments in the future.

Boyle broods upon the power of science and its possible abuse. He knows that some paths of inquiry should not be pursued even if possible, as in *a certain Quicksilver which I intend never to make againe, (and of which for the sake of mankind, I resolve never to teach the preparation)*. One's work in the lab can be dangerous, in the wrong hands discoveries can wreak destruction.

So here's someone worthy of emulation: diligent, scrupulous, not afraid to fail. But something of a workaholic.

In his own judgment, he had been bewitched. Even worse, all he had accomplished would seem primitive to those who followed—*they will be tempted to wonder why things to them so obvious, should lye so long concealed to us.* It was a painful realization, but that was the point—science meant moving on, leaving limited understanding like his behind. In the end he could make just one abiding claim for his life's work: *He made Conscience of great exactness in Experiments.*

Three hundred years later, scientists are still held to Boyle's stringent standards.

Consider "Dr. X." His work seems to constitute a dazzling leap in molecular electronics, and a prestigious journal has published his findings. But an alert colleague in another lab has noticed that something doesn't add up. Suspicion by itself isn't sufficient to settle the matter; a panel of dispassionate experts must be convened to go through the data and determine whether the discrepancies can be explained.

They start by positing good faith.

By all accounts, Dr. X is a hard-working and productive scientist. If valid, the work he and his coauthors report would represent a remarkable number of major breakthroughs.

Once the work is disinterred, however, a disturbing pattern of anomalies unfolds. In the early excitement no one had noticed that the physicist used duplicate graphs in different studies. And his results are consistently too neat, too precise, failing to reflect the fitful rhythms of a real experiment. He says it's not manipulation, just a series of honest mistakes.

His coauthors have clearly been duped themselves and are exonerated of scientific misconduct.

For Dr. X, however, exculpation is impossible and censure inescapable.

The investigative board concludes that he has committed the most heinous sin in science: *a reckless disregard for the sanctity of data.*

2

SCIENCE HAS OFTEN passed into popular perception as a kind of parlor trick. At the 1939 New York World's Fair, built around The World of Tomorrow, spectators poured into The Hall of Electrical Living to watch ten million volts of artificial electricity shoot through

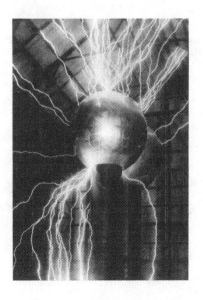

dazzling arcs thirty feet high, like a lightning bolt in your living room. Some of the fairgoers, in their youth, had seen the General Electric ads of 1899 promising that ELECTRICITY WILL SOON DO EVERYTHING. It was a miracle.

But scientists had been the first to marvel at the unexplained force.

Some, such as the Italian Luigi Galvani, professor of anatomy, furthered the work of pioneers in the field like Benjamin Franklin (who had called himself *an electrician*). But Galvani wasn't interested in kites or lightning rods; he wanted to tap into the energy that powers life.

He conducted public demonstrations in his Anatomical Amphitheater, where crowds watched as frog's legs and severed body parts—amputated legs and arms from accident victims, the decapitated head of a criminal—were probed with a wire hooked to nerves and muscles. Anyone could observe the violent twitches and spasms that ensued, the lifeless parts suddenly animated; one might conclude that every body carried within it a force Galvani called *animal electricity*.

He was hailed by contemporaries for *a discovery which does so much honor to all Italy.*

Still, he was cautious; *it was easy in experimentation to be deceived.* Techniques for studying the new force would have to be developed and shared with like-minded enthusiasts (*when sparks are to be sought, the experiment should be done in an extremely dark chamber*). But even a prudent man could be forgiven an occasional conjecture. Persons who are excitable, he reasoned, must suffer from *an excess of electricity.* It was conceivable that science could ease the burden of such disordered tempers, given the ever-widening understanding of the body's electrical fluids. And who was to say that this new field might not achieve something more profound, perhaps extending the very limits of life itself?

It was enough to keep a gifted man of science in his laboratory for a lifetime.

But science pays scant heed to individual aspirations. Incredulous crowds might not question Galvani's achievement, but a fellow scientist could challenge it. And so Alessandro Volta declared Galvani's work profoundly mistaken. He offered his own hypothesis: *Animal electricity* was nothing more than the charge occasioned by the moistened metal plates upon which the inert specimens rested. Touch them with a metal prod, as Galvani had done, and whatever was on the plates would convulse in response to the current. There was no resurrection of lifeless matter, just an agitated rush of what a later era would term *electrons*.

But that determination was still open to dispute in 1805, and the contested theories would war for ten years. The Società Italian delle Science was driven to intercede. It issued an open invitation to the scientific community: *Explain with clarity and dignity, and without offending anyone, the question of galvanism disputed by our worthy members.* Enthusiasts of electric fire took up the challenge throughout Europe.

The ensuing judgment of science was unsparing: Galvani had deceived himself, had seen only what he wanted to see. His theory would be relegated to the realm of failed science, along with such discredited phenomena as phlogiston. He would, however, bequeath the world a new way of describing excitability, as generations hence would be *galvanized*.

Volta's theory, triumphant, would be disseminated in his definitive *L'identità del fluido electrico* of 1814. He confirmed his thesis by building the first battery. It seemed itself miraculous; electrical power could be stored and drawn at will from a pile of metal plates.

But those who had dared to hope that science could vanquish death were disheartened to witness its diminution. What was a portable source of power to the promise of life everlasting?

But Galvani's work would live on, if only in the mind of Mary Shelley, who told the story of another broken scientist in *Frankenstein*.

One of the most startling images of the twentieth century can be dated to 1895. It is known as the *Radiograph of Frau Roentgen's hand* as taken by her husband, Wilhelm Roentgen, a German physicist.

For the picture, Frau Roentgen (born Anna Bertha Ludwig of Zurich) has been instructed to rest her left hand on a plate. In the subsequent image one cannot see any traces of her trembling, but she is susceptible to superstition and fears that her husband is invoking strange powers without the proper sanction or permission. (Even experts will express uneasiness; the 1901 Nobel Prize Roentgen is to receive for his work will allude to the *strange energy form* he discovered.) Roentgen has experimented with a variety of inanimate materials—wood, rubber, paper, foil, glass, lead—but now he needs a human subject.

Accordingly, against her better judgment, Frau Roentgen allows her husband to shoot a beam of high-powered particles into her body, where they move through her skin and flesh at unimaginable velocities until they hit bone. The resulting image is unlike anything she has seen before: In place of her hand there are now five slender skeletal fingers joined to a skeletal wrist. One of the fingers bears Frau Roentgen's

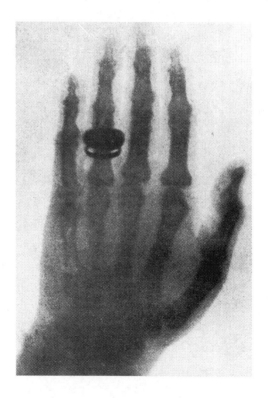

wedding ring, which like the bone has blocked the rays. Were it not for the ring, which she recognizes, she could almost hope that her husband had tricked her. But she cannot deny the truth: He has peered almost indecently into a living human body, exposing to view that which is better left to the imagination.

There's something altogether too intimate about it.

Dr. Roentgen, for his part, understands so little about the beams he's caught by accident that he labels them *X-rays*. He spends weeks experimenting with the phenomenon. He tries to contain his growing excitement: *I have discovered something interesting, but I do not know whether or not my observations are correct.* Though he works in solitude, he is building upon the investigations of Lenard, with techniques developed by Hittorf and Crookes, while using Ruhmkorff's coil—and all of these colleagues, he knows, worked themselves from earlier research. Thus he is only the most recent link in a long chain of implication.

In a sense, they are all responsible for this discovery.

When he is ready to announce his findings to the world, Roentgen nervously mails the extraordinary images to a small circle of experts. He worries that he might be considered unserious, even a crank, by his colleagues. And his wife's alarm, while unsophisticated, still bothers him. But there is no turning back, as he notes—*Und nun ging der Teufel los* (and now the Devil was let loose). His official report on the new rays, *Über eine neue Art von Strahlen,* is delivered to the Würzburg Physico-Medical Society just after Christmas.

His apprehensions prove groundless; indeed, he finds himself a suddenly famous figure in a world that has been waiting for revelation.

Perhaps he had forgotten that mankind has always been drawn to the Devil. In just a year, over a thousand papers will be published on the X-ray as researchers rush to explore the phenomenon. Some, ambitiously, will attempt to capture the human soul on the photographic plates. An enraptured public will embrace the discovery, transfixed by its powers of transparency.

In America, the brilliant loner Nikola Tesla boosts the power of Roentgen's rays tenfold in his own lab. He lectures the New York Academy of Science on the peculiar appeal of the ray: *the desire to see things which seem forever hidden from sight is more or less strongly developed in every human being.* In his laboratory records he notes the curious blistering of his skin that seems to accompany the experiments, but also another unexplained effect: *It is a fact that the troublesome cough with which I was constantly afflicted has entirely disappeared.*

The data is accumulating, bit by puzzling bit.

H.G. Wells stokes the fascination among laymen with *The Invisible Man*, a cautionary tale published two years after Roentgen's announcement. The Invisible Man is a scientist who has *meddled in things men should leave alone,* according to a timorous colleague. *Straightforward scientists have no need for barred doors or drawn blinds.*

Dr. Griffin tells us he began his scientific career as a struggling chemist who aspired to greatness. He spends every spare moment in his lab, monitoring the noxious brews he pours almost reverently into glass vessels. After *a thousand experiments, a thousand failures,* he has succeeded; the formula for invisibility will certainly place him in the scientific pantheon. But more rigorous tests must be conducted. At this level of inquiry, with risks that border on recklessness, only the scientist himself can be considered a suitable subject. As he drinks his potion he disappears, flesh dissolving layer by layer. For a time, invisibility is immensely diverting.

But dementia eventually sets in, a possible effect of unseemly scientific ambition. He is driven to great crimes while those closest to him, his few lingering loved ones, comb through his inscrutable notes with pained incomprehension. In his rational moments Griffin almost wishes he had never initiated his experiments. But the process is irreversible; even science cannot put a genie back into the bottle. As a consolation, his torment has warned others away, and his work dies with him.

In the real world, those with fewer qualms are pressing forward. One obscure woman labors with her husband in a makeshift Parisian workshop they've built in a shed. She has decided to investigate Becquerel rays, one of the myriad *invisible radiations* that she senses in the very air around her. Unlike Frau Roentgen, however, Madame Curie embraces the unknown; the mysteries of the natural world she finds almost seductive, though she would blush to speak of her work in that way. In any case, she regards those prone to superstition, including religious belief, as simply persons who possess *a limited state of mind*. Sometimes, she envies such easy certitude.

In the shed she is attempting to isolate the source of a new phenomenon she has termed *radioactivity*. But the force won't readily reveal itself; it must be coaxed from the huge iron pots filled with bubbling chunks of rock and ore. She stirs the simmering concoction with an iron rod until her arms ache. It will take four years in this chemical kitchen to reduce the tons of ore to one gram of the substance she dubs *radium*.

She has identified an entirely new element and revolutionized the research into subatomic particles. In a sense, she (like Roentgen) has invented the twentieth century. She will become the first woman to receive a Nobel Prize for her work (her scientist daughter will become the second). Her home cookbooks will remain radioactive for centuries, tokens of competing passions.

But it's the wonder itself, she knows, that will last the longest: *Whenever some chance discovery extends the limits of our knowledge we are filled with amazement.*

3

DECEMBER 17, 1903, 10:35 AM, Kitty Hawk, North
Carolina: The world's first powered flight of an aircraft
lasts all of twelve seconds, but one of the five eyewitnesses on hand will proclaim, years later, *it was one of
the grandest sights, if not the grandest sight, of my life.*

For millennia, men have attempted to fly and have failed. Those who believed in the possibility of a *man-carrying machine* have been mocked. But two plainspoken brothers from Ohio, Wilbur and Orville Wright, have taken up the challenge.

They begin to read up on what they call *the soaring problem,* working their way through books from the Dayton Public Library. (Their home bookcase already includes volumes such as *Fragments of Science* and *Up the Heights of Fame and Fortune.*) Wilbur purchases additional texts from the distant Smithsonian Institution in Washington DC. As they talk, and study the latest thinking on the subject, a thought of their own begins

to grow. As Orville will later put it: *One day we said to each other, why not?*

Their obsessive fascination with flight began in childhood, with a toy their father brought home in 1878. He told them it was a *helicoptere* and tossed it into the air, where it whizzed about for a while before flying into the ceiling and spiraling to the floor. The boys were intrigued; nothing they played with on the ground had ever seemed so interesting.

There will always be something childlike about them; as adults, they make their living repairing bicycles. One hundred years later, the Smithsonian will carefully preserve a claim check from the Wright Cycle Shop, divining it for clues to the mystery of their development. It hints at an early reliance on procedure: *we guarantee all work, but complaints must be made properly.*

Neither man will marry. The after-hours diversion that started out as sport will come to consume them, claiming the rest of their lives. In retrospect, they won't know much more than anyone else about how it happened, except for a feeling that a force had taken hold of them: *We were drawn into it deeper and deeper.*

History will hardly be able to distinguish the two.

Early on, they blur into one being in their own minds: *we lived together, played together, worked together and thought together*, one of them famously remarked to a reporter.

But research does reveal some distinctions. Orville is the excitable one, younger by four years. He peppers his letters with exclamation points—*we have found out again that everybody but ourselves are very badly mistaken!!*—and possesses something of a dramatic streak: *P.S. Please do not mention the fact of our building a power machine to anybody.* He lies in bed at night imagining the flying machine, too distracted to sleep, thrilled at the thought that he and his brother might succeed where everyone before them has failed. He's the one who writes dutiful notes to the family back in Dayton when they're camped out at their research site (*Rations are getting low again . . . we are running a fire all night now to keep warm*). Sometimes, he runs out of steam and simply stops (*I can't think of anything else to say, so good-bye*). How can he possibly communicate their adventure to anyone else?

Wilbur is the more natural experimenter of the two (or, more precisely, *aeronautical engineer*, though that useful term is yet to be invented). He invests his faith in exactitude; if they harness the power of calculation they cannot help but prevail. But he battles a melancholic temper that infuses even his business correspondence —*for some years I have been afflicted with the belief that flight is possible to man.* He seems to be ambivalent about his ambition; perhaps it's the intensity that makes him uncomfortable.

He parses his language as carefully as possible. *I am an enthusiast*, he writes, *but not a crank.* And he wonders how others can use words so loosely; what good does it do to speak of an *angle of incidence* if the term is applied elastically? He proposes a rule for general reference: *the angle of incidence is fixed by area, weight, and speed alone.* If an experimenter balks at the formulation, he probably shouldn't be pursuing the flying endeavor at all.

Of course the Wrights have certain advantages over others. They have a loving family around them, run by their elder sister Katherine since their mother's death. She attempts to be encouraging, though it doesn't come naturally to her. She's mystified when the boys go off to work on a remote North Carolina beach hundreds of miles away. *I never did hear of such an out-of-the-way place.* Occasionally she complains further to her father, who travels, in comments tinged with mild exasperation.

We don't hear anything but "flying machine" and "engine" from morning to night. It must be difficult for her, listening as her brothers talk to one another in their own odd language, watching as they hunch over their worktables with the grave intensity of schoolboys. Her mother might have been more comfortable with the notion; she used to make simple toys for the children and sometimes kitchen aids for herself. But she was inventive, while Katherine is simply capable.

Wilbur and Orville agree that they will advance one step at a time, taking the glider up over a thousand times before even thinking of anything more: *If we find it under satisfactory control in flight, we will proceed to mount a motor.*

They study birds; they discern telling differences in the flying styles of bald eagles, ospreys, hawks, and buzzards. They are not the first to look to ornithology for inspiration (Leonardo designed wings for a man, and the great glider specialist Herr Lilienthal—whose untimely death spurred the brothers on—wrote the classic *Bird Flight: The Basis of the Flying Art* in 1891). But only the Wrights (or more precisely, Wilbur) have determined that *the buzzard weighs about .8 pounds per square foot of wing area.*

Thank goodness for adverbs like *about*.

He and Orville build hundreds of wing surfaces, shaved and sanded to minutely varying specifications, testing each scale model in a small wind tunnel that is also homemade. The tiny wings are just a few inches long, like the toys they took apart in childhood. Sometimes Wilbur will upbraid his brother, who should know better than to move while the wind tunnel is in operation, disturbing the airflow in the room and possibly affecting the accuracy of the instrument. Eventually, the results they achieve will vary by just one-tenth of a degree.

Along the way, as each additional component of the machine increases its complexity, they encounter one setback after another. It's a textbook case of discouragement. Their father keeps an account of their travails: *the boys broke their little gas motor in the afternoon.*

He must wonder where it will all end.

It will take years of observation, calculation, testing and retesting before the Wrights file U.S. Patent 821,393 for *a flying machine*. Wilbur will give greatest credit to the data they have developed, entry by entry, penciled-in tables proliferating until eventually the reference books have to be rewritten. The scientist in him knows that they owe everything to the correct reckoning of the numbers. He develops another formulation for posterity: *Sometimes the non-glamorous lab work is absolutely crucial to the success of a project.*

Ultimately, in the annals of science, the Wrights will be celebrated for the methodical techniques they've developed as much as for the achievement of that December day.

In 1908 Orville will pilot the first passenger flight at the invitation of the Army, but an unfortunate mishap that morning will result in the first fatality of powered flight. His passenger, Lt. Thomas E. Selfridge, will earn that statistical distinction. He probably won't have time to ponder, as the craft goes down, that he was a last-minute substitute for a previously scheduled passenger.

Orville, the happy-go-lucky one, will survive to fly again, and outlive his brooding brother Wilbur by thirty-six years.

In time the *flying machine* will become the *aeroplane*,

and students in elementary schools will be asked to write essays imagining what life would be like without flight. By 1930 a typical child will be able to observe: *Up, up in the air / I see airplanes flying everywhere.*

And one day, in the future, the command module of a manned flight to the moon will be christened *Kitty Hawk.*

4

ONE CAN'T JUST read about science; at some point, one ought to get a little hands-on experience.

But how to start? Here's a thought—why not take a ten-minute field trip?

The idea is the inspired strategy of Helen Ross Russell, a science educator in the 1970s who believes that the scientific mind ought to be molded early in life. Not every school has a lab, or even a science program. But if she can get fidgety students in an urban setting to carry pencil and paper around for a few minutes at a time, the basic skills of measuring and recording might become second nature as they explore their schoolyard. These ten-minute trips might whet their interest in the world, and who knows what unanticipated achievements might one day result?

But how can a token immersion lead to serious science? A little research reveals precedents for such an approach.

The twenty-two-year-old Charles Darwin, as a junior naturalist, spent five long years traveling around the world on his first major expedition. But he had taken more tentative forays earlier, beginning with short walks around Wales, observing the landscape closest at hand. His teachers had insisted that his training begin modestly.

Maybe Russell is on to something.

I keep on steadily collecting every sort of fact, the young naturalist noted, already observing himself with detachment. That's an attitude one can learn to adopt anywhere—a scientific mindset becomes more pronounced with practice.

As a bachelor he had charted the pros and cons of marriage in a notebook, recognizing qualities in his wife-to-be that he himself did not possess: *I think you will humanize me, & teach me there is greater happiness, than building theories, & accumulating facts in silence & solitude.*

But an older Darwin instructed his son Horace, an aspiring scientist, that scientific discovery required *habitually searching for the causes & meaning of everything*, a task that required complete commitment. Darwin, better than anyone, knew how easily such investigations could ensnare one. As a young man he had intended to spend just a few months thinking about barnacles (*Thyrostraca, Cirrhopoda, Cirrhipedia*), but found himself absorbed in their study for eight years.

At that rate, a whole lifetime can pass in what seems like a nanosecond.

Maybe childhood is indeed the time to instill a feel for the joys and stringencies of science, before the possibility of regret sets in, before one can apply a proper risk-reward calculus.

By the time Darwin wrote his *Autobiography*, after he'd achieved worldwide recognition, he had become a master of dispassionate regard: *I have attempted to write the following account of myself as if I were a dead man in another world looking back at my own life.*

Whatever qualms he'd once felt had long since been subordinated to the encompassing claims of science.

Maybe it's possible for a nonexpert to build *a simple machine*. That can't be complicated, there seem to be only six: wedges, levers, screws, inclined planes, pulleys, and the wheel and axle. It's not string theory, after all; ordinary people have been making and using these machines for thousands of years.

But the definitions themselves are in dispute: Some argue that a wheel is essentially a lever, while others hold that a wedge is actually an inclined plane.

Without agreement on the basics, how can one move on to anything more complicated?

Why not invent something?

Apparently even a high school dropout can do it, like Florence Melton. She left school in 1929 to help support her family by working in a department store.

Soon enough it's 1947, and she's reading *Popular Mechanics* in her spare time. One of the articles catches her eye—an announcement of a new material, a foam made out of rubber by a company called Firestone, something developed for the boys overseas to help cushion their helmets.

Melton is one of those people who see possibilities in the unlikeliest of places. You can't teach that in school.

She grabs Aaron, her husband, and they're off to the Firestone plant in Akron, Ohio, where she talks to the experts, finds out whether a woman's shoulder pads (the family business) can be manufactured out of foam. It would be so much easier on everybody. Convenience, like the idea of the consumer, is another notion that's new.

It's an unusual request, but there don't seem to be any technical limitations, and everyone benefits from new business. The Meltons sign a contract for the product and head home.

He's watching the road, maybe wishing he had cruise control for the long drive back (it's just been introduced). But she's thinking. All those years in retail, on her feet all day. He can't yet see it, but for Florence the future has just swung into full view.

Aaron, you know what we ought to do with foam rubber? We ought to walk on it.

Over the next thirty years their company will sell a billion pairs of what she calls *slippers*.

Why not patent something? Something that people will use, and don't yet know they need.

Such a revolution in everyday life has often occurred as a happy offshoot of science.

Patent law requires that an invention *not be obvious*, meaning that someone with ordinary skill wouldn't think to come up with a similar solution to a perceived problem or need. Otherwise, everyone would claim the credit.

There may be no better model of such ingenuity than Ruth Siems, awarded U.S. Patent 3,870,803 in 1975.

The problem: How to devise a bread crumb with a certain particle size that possesses a firm cellular structure such that it hydrates properly in less than ten minutes.

The solution: Stove Top Stuffing.

It seems obvious enough in retrospect, with sixty million boxes of the mixture sold every Thanksgiving, but until Siems invented Stove Top no homemaker would have thought to serve an instant stuffing to family or guests. Traditional food took forever to cook, and women were resigned to remaining in the kitchen while everyone else relaxed.

Siems had begun her career at General Foods in cake mixes and flours. She knew that a successful stuffing would be devised only by someone with a knack for food chemistry, someone who cared about texture, taste, color, and mouth-feel. It would all come down to the crumb.

Eventually, after a period of trial and error, she came up with an equation of sorts: *said crumb having a particle size such that at least 95% by weight passes through a 2 mesh screen and no greater than 5% by weight passes through a 50 mesh screen.*

Later in life, she shrugged off the achievement. *I've always liked to put things together.*

Given such evidence, it might be a propitious moment to dismiss the canard that women can't do science (in one form or another, the claim has been made for hundreds of years). Consider the three young ladies who won highest honors in a 2007 national competition in Math, Science, and Technology.

It's the first time that girls have won the prestigious prizes.

Isha Himani Jain, sixteen years old, devised a study of bone growth in zebra fish, a process that could shed light on adolescent skeletal development.

Janelle Schlossberger and Amanda Marinoff, high school seniors, created a molecule that helps block the reproduction of drug-resistant tuberculosis bacteria.

Alicia Darnell did research identifying genetic defects related to Lou Gehrig's disease.

Studies do show that girls sometimes get discouraged if they experience a setback, questioning their abilities, whereas boys tend to dig in and redouble their efforts. But tenacity isn't a gender-based trait, and it may be more readily transmissible than we imagine.

Each of these young women shows unusual promise in science. And maybe they've got little sisters.

Further inspiration for girls can be found in the more distant historical record. Consider Lady Ada Lovelace, born 1815, daughter of the volatile Lord Byron. Her mother had forced her into scientific studies, desiring to expunge any trace of the poetical frenzies of her father.

He no longer lived with them, but precautions must still be taken.

Lady Byron believed that rational studies offered the surest antidote to the extremes of the poetical temperament. And, in any case, the profligate had expressed disappointment at not getting a son. He didn't deserve emulation.

Ada would take to her studies and, in fact, become a mathematical genius, one of the few who could understand Charles Babbage's design for his Analytical Engine. She compiled additional routines for the device, now considered the world's first computer.

She enthusiastically translated the few articles written about it for the scientific elite of Europe.

Babbage wondered why she didn't just write her own essay on his invention. Apart from himself, she seemed to know it best. Lovelace had no answer, except to allow that such a thought had never occurred to her.

Ironically, shortly after Ada's birth, Lord Byron would encourage Mary Shelley to write *Frankenstein*. He knew that passion was no stranger to science.

And in the 1970s, a new computer programming language would be named *Ada*, in honor of the woman now recognized as the first computer programmer.

Scientific careers don't always develop steadily; sometimes, they begin with a radiant burst of imagination or luck, you might say they go supernova at the start. Then, just as suddenly, they flame out.

It's a rare occurrence, nothing that most of us can ever hope to match.

Take Stanley Miller, a twenty-three-year-old graduate student searching for a thesis project in 1953. His advisor, a Nobel laureate, is known for his research into the origins of the solar system. The related science of cosmochemistry has been bubbling fitfully along since the 1920s, when Oparin and Haldane first advanced the notion of *primordial soup,* a kind of oozing swirl from which life is thought to have emerged. Or may have, no one can say for sure, it's just an idea. It hasn't been tested. Miller considers himself a theory man and resists experimental work—it takes too much time, for one thing—but he'll be a graduate student forever if he doesn't find something to do.

What about this idea of cosmic soup? Why not build a model of the early Earth and see what he can cook up? He won't need detail—no landmasses or little people—so the execution should be simple. And it might be fun to create life in the lab.

His advisor doesn't think he's ready. It's just like a young man to get ahead of himself. He should do something more mundane, something with meteorites. But Miller won't let it go, he's already designed the apparatus.

A glassblower constructs the mechanism, two large clear bubbles connected to glass pipes and amber flasks that look like Molotov cocktails. Electrodes will be discharged into the upper sphere, simulating the effects of

lightning bolts above the ocean (that's the water slosh-
ing in the lower sphere). Miller has developed a recipe
—methane, ammonia, a touch of hydrogen—meant
to mirror the noxious gases in the Earth's early atmo-
sphere. If he's right, primitive organic compounds, the
building blocks of life, ought to appear.

He lets the mix sit on low heat for two days. He can
see something purplish, a spot, but the impression is
faint. It might be an amino acid, it might be a mirage.
He's got to accelerate the process somehow.

Why not crank it up to a boil?

Science will record that this impulsive action led to

Miller's eureka moment, an exultation first described by the Greek Archimedes in antiquity. It involves exuberance, a sense of disorientation upon a discovery, and it makes scientists unnaturally giddy.

He was three feet off the floor, Miller's brother recalled decades later.

Everyone from Enrico Fermi to a young Carl Sagan hailed the news. Miller had done something almost miraculous. He'd get his degree now.

Late in his life, some forty years after the experiment, he will still be posing for pictures alongside his aging apparatus. By that time, the original theory will have suffered; alternate conjectures about the origin of life have found some favor.

But no one who was there will forget that early supernova.

5

NEWTON, FARADAY, EINSTEIN, Watson, Crick—one could easily study only the illustrious, as if science were an enterprise undertaken only by those whose names are already well-known. There's a term for it—hagiographic science—a process whereby one fetishizes the scientific elite, if inadvertently, and ignores the legions of capable practitioners who work in utter anonymity.

That's certainly a tendency one sees outside science. You might call it *the big shot syndrome.*

Scientists themselves adopt the idea of a pecking order early on; generations of faceless graduate assistants have willingly labored to someone else's greater glory in a lab. It's sacrifice for its own sake, a professional rite of passage. Get through that, and you can pass the lab coat on to some other shlub.

Recent studies theorize about the underachiever

phenomenon in science as well. Simply put, it may be that scientists with low productivity are just *second-rate researchers*. End of story. But to a nonscientist the unsung, as it were, seem to have earned their lack of distinction, if only by flouting the field's norms and doing very little, very well. One has to admire the counterintuitive outcome.

In an odd way, it's a wash.

Even reputations of the eminent sometimes deflate over time. It's unsettling to learn, for example, that many contemporary scientists believe the great Marie Curie was overrated.

Still, at least she had her shot.

Who remembers Alfred Russel Wallace? Who knows that he formulated a theory of evolution independently of Darwin? In a moment of innocent enthusiasm, the fateful act of a naïf, he sent his detailed ideas to the more established scientist. His agitated elder was suddenly spurred to put his own findings in order.

Darwin's book came out first. And he took pains to make it accessible to ordinary people.

Wallace was gracious enough to accede to the inevitable as acclaim slipped away and settled on someone else. He delivered lectures on "The Darwinian Theory" to excited audiences in America. He knew that his own life's work had been eclipsed by a twist of fate, that his twenty-one books and hundreds of articles would hardly become more than historical curios, footnotes to the titanic *Origin of Species*.

Even during his lifetime the loss of scientific gravitas was apparent. *He has the bearing of an ordinary citizen rather than that of a scientist*, the *Boston Herald* wrote of Wallace in 1886, in an early use of an informal metric that would come to measure the gulf between experts and everybody else. If you look smart, or seem to be aloof, you must be a scientist; if you appear to be approachable, or even slow-witted, you're more likely to be a layman.

It's perception that matters in the public mind.

The ultimate historical verdict on Wallace would be briskly matter-of-fact, acknowledging that *he has all but vanished from public consciousness.*

Another question: Who was the second man to walk on the moon? It's the kind of challenge that might stump a player in a parlor game. You'd win points if you knew that the answer is Edwin "Buzz" Aldrin.

He had taken the same life-rending risks as Neil Armstrong, descended to the lunar surface in the same fragile two-man craft that no one had yet tested in space. He could have been killed in the service of a national obsession. One would think that being number one wouldn't be so important in this instance, the two trips down that ladder separated by just forty minutes.

But that isn't the way it works.

In a single moment Armstrong became the most well-known man on Earth. And Aldrin wasn't like Wallace; it would take him years to make peace with the second-class hand that history dealt him. Thirty years after the flight, the oblivion would be all but complete, as even the memorabilia associated with his role in the moon shot would languish from lack of interest. ALDRIN'S MEDALLION FAILS TO SELL ON EBAY. The headline would make painfully clear what Aldrin himself had long known; the public had already moved on, before he even returned to Earth.

6

LET'S TRY a "thought experiment": It's one of the most useful tools that scientists have devised to grapple with perplexities. One doesn't need a large, well-equipped lab, funding, or even colleagues—just an imaginary set of circumstances and a hypothesis. In effect, it's a process of reducing problems to essentials, and working them through in your mind.

Galileo and Einstein did it; later, Putnam famously imagined the twin earth scenario, and Searle the Chinese room.

In theory, anyone can devise something similar.

⊘ Thought Experiment

Assume that you have very little time to learn as much as you can about science. Imagine, if it helps, a doomsdaylike device that ticks loudly as you study, with a digital readout that counts down to zero.

Assume further that you possess at least average intelligence. You must proceed in spite of the knowledge that even the American Association for the Advancement of Science thinks the odds are against you—according to that august group, the gold standard for such speculation, there is simply too much scientific knowledge for anyone to acquire in a lifetime.

So far, only a fool wouldn't be worrying.

Assume further that you may be one of those many adults who fail to grasp the most elementary items of scientific knowledge. You may not realize that the Earth revolves around the Sun; you may think that humans and dinosaurs lived at the same time. You will almost certainly not be one of the tiny minority *(in the words of one post-*Sputnik *study) who see science as exciting.*

Question: How much material could you master, and how quickly? Is it worth pursuing such knowledge, in the face of almost certain failure?

Remember: The clock is ticking.

London, 1771. The first edition of the magisterial *Encyclopædia Britannica* is published, its purpose *to diffuse the knowledge of science.* Its editors have pored over the finest books written upon every conceivable subject, distilling (in three volumes) the latest thinking of a society bent on propelling itself into the future.

The *Britannica* inveighs with special fervor against *imaginary causes*, those fantastical notions that the misguided, clinging to archaic conceptions, have seized upon to explain the most puzzling aspects of nature.

One so-called substance is singled out for particular scorn.

Aether is the name of an imaginary fluid, supposed by several authors, both ancient and modern, to be the cause of gravity, heat, light, muscular motion, sensation, and in a word, of every phenomenon in nature. We are sorry to find that some late attempts have been made to revive this doctrine of aether.

One can only pity those unwilling to give up the ghost.

A new way of thinking about the world has arisen: Knowledge is to be derived from experiment and observation, not wild speculation. The latter leads only to *folly*, a state against which the thinking man must ever be on guard.

○ ○ ○

Certain subjects, in 1771, can safely be considered settled, among them the design and constitution of The Ark—*a floating vessel built by Noah, for the preservation of his family, and the several species of animals, during the deluge.* The Ark lends itself superbly to the new emphasis upon calculation; Moses himself has supplied the numbers, and the *Britannica* is able to declare its dimensions three hundred cubits in length, fifty in breadth, and thirty in height, *which some have thought too scanty, considering the number of things it was to contain.*

But the new science affords the tools to settle such disputes.

Buteo and Kircher have proved geometrically, that, taking the common cubit of a foot and a half, the ark was abundantly sufficient for all the animals supposed to be lodged in it.

Providentially, there was also room for the thirty or forty pounds of hay that each ordinary ox would consume in a day. Even a man of faith, like the learned Bishop Wilkins, marvels at the melding of new ways with old: *The most expert mathematician at this day could not assign the proportion of a vessel better accommodated to the purpose.*

The curious reader who edges further into the *A*'s might happen upon *Astronomy*. There are six major celestial bodies in the solar system, each circling the sun, the most imposing of which is *Saturn, the outermost of our planets.* Its magnitude is such that no one who looks upon it can imagine it destitute of rational creatures. But the eighteenth century isn't equipped to explore the cosmos, save at a distance, and in any case a benevolent Creator has given the man of science a profusion of riches right here on Earth. Those of his creatures flourishing on Saturn will have to study themselves.

And so it goes, *A* to *Z*, crystalline nuggets of knowledge suspended in a moment in time. Reference works of the future won't be so entombed; knowledge will morph along the *Wikipedia* model, evolving moment to moment in a kind of ceaseless churning so that nothing is ever quite settled again.

☉ *Thought Experiment*

Imagine you're a planet, a junior member of the solar system. You're not much to look at—just a runt, truth be told—a sphere of icy methane 2,274 kilometers in diameter. That's even smaller than Earth's moon. But you're spunky. It takes you 248 years to orbit the Sun, yet you've done the job for eons without complaint, making your distant rounds on schedule, never expecting a medal just for showing up.

You're almost never visited—you see only the occasional NASA probe—and the next one's not even due until 2015. They're more interested in the big boys like Mars, Jupiter, and Saturn. Even Neptune, your neighbor, doesn't keep in touch. Still, you hang in there. It's a big universe, there ought to be room for everybody.

One day, you're given your pink slip. Just like that, after seventy-six years of service.

You're no longer a planet: You've been downsized to a dwarf. You get the usual spiel about how times have changed, they've got more powerful telescopes, they're finding objects bigger than you all the time. Something's got to give, otherwise we'd have ten, maybe fifteen planets, and no child can be expected to memorize that long a string in school. Yada yada.

Question: Do you accept your expulsion, gracefully, or do you chant "My Very Enormous Monster Just Sucked Up Nine Planets" to anyone willing to listen?

7

Lindberg & Plane Over City. It's an old picture, carefully preserved, captioned with the white ink that no one has used for at least half a century. It looks like it was taken by an amateur, hanging out a window, a tourist trying to capture a moment of history with a home

camera. Another hazy image of the past that survives in a faded family album.

There's the city, for sure, but where's Lindbergh? All one can see for certain is empty sky, and some small dark specks that are just as likely to be dust motes as an airplane in midair. Maybe the documentarian has gotten ahead of himself, pulled the trigger a second too soon, or a moment too late. But early photographic equipment is bound to be primitive; the whole point of the picture has got to be there somewhere.

And so it is, slightly off-center, a small, ghostly bird sailing lightly through the air.

Yes, it's very likely a picture of Charles Lindbergh attempting the first solo, nonstop flight across the Atlantic Ocean on May 20, 1927.

The $25,000 prize for the feat was first announced in 1919, but every pilot who's attempted the transatlantic flight to date has been killed, injured, or has disappeared. It's as if the Atlantic has become a graveyard of ambition. No one in a later age will understand how dangerous the flight seems in 1927, how crazy you'd have to be just to attempt it, how easy it is to lose one's way over 3,600 miles of ocean.

It's not a stunt like eating goldfish, after all.

But Lindbergh's been thinking about the challenge for a while now. He's already fallen for flying, the way young men in the future will fall for cars and chrome. He's impressed with the new Wright Whirlwind J-5C, the first engine that might reliably power such a long-range trip. It's been tested to 150 hours or so at a clip, several times the expected duration of the flight. With that kind of leeway one can even make a mistake and hope to recover.

He doesn't care about the money. He may not even know why he really wants to do it. He certainly doesn't yet understand that it will prove the defining undertaking of his life.

His simple bearing, the unassuming manner of an especially laconic Midwesterner, endears him to a public that hardly knew him just a week earlier.

Everybody knows him now.

Newspapers ask that steamship captains at sea and sentinels on land watch for the call letters N-X-211 on the wing of any plane flying overhead.

They want to protect this slip of a boy who probably thinks he's invincible.

◦ ◦ ◦

He built the plane itself with an engineer, *every part designed for a single purpose*. He considered all the contingencies he could imagine, beginning with the basics: *1. Successful completion; 2. Complete failure*. He consulted texts on navigation, and taught himself spherical mathematics to plot a course in hundred-mile increments across the Atlantic.

He knows every inch of the map. He's ready for any situation within expectation; anything else is simply out of his control. *Such hazards aren't marked on the best maps one can buy.*

Equations are easier to handle than the inchoate emotions that sometimes overwhelm him. *The symbols I pluck from paper, applied to the card of a compass, will take me to any acre on the earth I choose to go*. All he needs are the numbers; self-analysis is for some other kind of person. Accuracy, he declares, is *vital to my sense of values.*

He takes no radio, just a raft and four flares should his plane go down somewhere over the Atlantic. He'll take a stab at staying alive if things don't work out, as much for his friends as for himself.

He knows that most people don't understand why he wants to go it alone—maybe he doesn't need a co-pilot, but what about company?

They will never be drawn, as he is, to the seductions of solitude: *Like the moon, I can fly on forever through space.*

The *Spirit of Saint Louis* itself is claustrophobic, just a wicker chair in a small cabin. He's taken the front window out to store more fuel, so he sees only the instrument panel as he looks ahead. He'll have to tolerate complete physical immobility for up to forty hours— he won't be able to stretch his legs or sleep. At times, fog will envelop his craft so completely that he won't be able to see anything at all, and when he does see the ocean, below, it will have spread out like a dark cloak that hides the world from view.

He fights fatigue, resisting the jumble of his own thoughts as he begins to tire. He treats his mind like a machine—*it must be kept on its proper heading as accurately as the compass.* Late in the trip, disoriented, he turns away from the European coastline and heads back out over the ocean, but he has an instinct for sensing error and turns around in time.

In a dark period midflight he'll have thoughts that would never have occurred to him on land. (*Is this a dream of death I'm passing through?*) But a poet's sensibility won't keep him on course; feelings must be subordinated to fact. *Science is truth, science is knowledge, science is power,* and that power will keep him alive. It's a lesson he learned from his grandfather, and he's never forgotten it.

The historic 3,600-mile flight will finally end at Le Bourget Airfield in Paris, 33 hours, 30 minutes, 29.8 seconds after taking off from New York.

After the flight, he'll be called *Lucky Lindy* by a public that doesn't understand how little luck had to do with it.

In 1928 Lindbergh will give his beloved *Spirit of Saint Louis* to the Smithsonian Institution in Washington DC, where it will hang, a little lonely, long after its pilot's life has ended.

But in 1927 he's still a young man; how can the future be anything but anticlimactic?

He will never prove especially fit for the regularities of what he calls the *antlike life*, though it will have its moments of upheaval, as when his firstborn son is kidnapped and killed just five years after the flight.

That will only drive him further into himself. But the trajectory of his withdrawal from the world was set from the moment he stepped into his plane in New York. It had only been thirty-three hours, but he knew he would never feel such freedom again.

Forty-two years after his flight, Lindbergh himself will experience the wonder of witnessing a spectacular feat of aviation. Later, he'll write a note to Michael Collins, the only crewmember of Apollo 11 who did not descend to the lunar surface:

There is a quality of aloneness that those who have not experienced it cannot know—in some ways I felt closer to you in orbit than to your fellow astronauts I watched walking on the surface of the moon.

He had waited most of a lifetime for someone else to understand.

8

◎ Thought Experiment

Imagine you're a twentieth-century painter who was once linked romantically to both a world famous artist and a pioneering scientist. (It's irrelevant, in this scenario, that you were the only woman with enough sense to leave the artist before he left you.) Assume that each of your partners was widely and justly recognized in his time as possessing unusual imaginative powers.

Assume further that you've been entrusted with the legacies of each relationship. You revere the practice of art, but you also have great regard for the achievements of science.

One day, a fire breaks out in the warehouse where you've stored the works of each man. You don't have time to rescue everything.

Question: Which would you choose to save—a number of profoundly important paintings or a lifesaving vaccine?

◎ ◎ ◎

The French countryside, 1940: Schoolboys have stumbled upon a cave, deep in an area where no one has ventured in living memory. At first, the boys see nothing in the dim light as they work their way into the space, but the walls come to life as they look more closely. There's an animal in midgallop, alongside other brilliantly colored creatures.

The images are strange but thrilling.

Local villagers flock to the site, and soon the experts arrive. Everyone points in awe at the paintings. Archaeologists surmise that the images are seventeen thousand years old, and arguably *the most remarkable Paleolithic cave paintings in the world.*

In the wild, bright pictures of bison and bulls one can see the touch of an early Picasso, perhaps. One wonders if the magnificence of the work was apparent to his peers, or if some questioned even then whether it was really art.

It's tempting to search for connections across the centuries.

Over two thousand images, with just a scattering of Homo sapiens. In one sequence a stick figure falls backwards, gored by a huge animal with horns. The modern conservators have titled it, helpfully, *Scene of the Dead Man*. It may be evidence of an early understanding of cause-and-effect. Or a newfound instinct for narrative—maybe the fellow who's fallen wasn't much of a hunter, and this is a cautionary tale. A kind of Darwinian warning, before humanity had a word for it. As to the figuration itself—why would the animal be rendered so much more realistically than the human? It's as if a child had begun the drawing, and an adult had obligingly finished it.

Contemporary experts tell us that questions arising as one studies such primitive paintings will never be answered; *they were certainly created for a purpose, but today we can only guess at the ideas or emotions they were meant to convey.*

These Ice Age hominids seem unimaginably remote.

Public interest in the caves grows; people want to see the site for themselves. But by the early 1960s it's apparent that something's gone terribly wrong, the pigments are fading, images that have lasted millennia seem to be disintegrating.

It's only been twenty-some years since their unveiling. What could have changed?

Once you put it that way, start isolating the variables, the answer to the anguished question is almost obvious. Anyone can calculate it: Twelve hundred sightseers per day, so many exhalations per minute, it must be the carbon dioxide that's doing it.

And once the cause is established, it's just a matter of reversing the conditions that led to the corrosion.

That means, of course, keeping the public away.

From that point on the site will be monitored, like a sickly patient on life support. The Laboratoire de recherche des monuments historiques has devised the protocols. Only minute deviations in environmental tolerances will be allowed. The site, restored, will be sealed, the startling images entombed once more.

No one but scientists will ever see the originals again.

But the public is getting used to the idea of reproductions. They buy pictures of paintings now, why not just make copies of the images, perhaps in situ? Science can duplicate the site with great exactness—there's even a method of measurement that will re-create the uneven curvatures of the walls. The images themselves can be projected onto this artificial surface and then painted in, as with a giant color-by-numbers kit. The new site can be called Lascaux II, to make sure no one is misled.

And so the replica is created just down the way from the original venue. Someday (perhaps seventeen thousand years from now) experts may avidly examine the work of two primitive cultures.

But it's important to keep one's perspective.

Sign in a prominent science museum: THE PAST 30,000 YEARS OF HUMAN HISTORY SPAN THE WIDTH OF A SINGLE HAIR AT THE END OF THIS COSMIC PATHWAY. The pathway itself is about a hundred feet long. It begins with a marker at the Big Bang (14.5 billion years ago) and continues to the present. The single strand of hair is encased in Plexiglas. There's no identifying DNA; it could have been donated by anybody.

The implication is inescapable: The Lascaux painters and their twentieth-century counterparts coexist in essentially the same moment in time, sentient life dwarfed by the emptiness of eons.

9

◎ Thought Experiment

Imagine a moribund world where nothing stirs. There is no wind, no water, no vegetation, and certainly nothing that we, in our limited experience of the universe, might recognize as life. There is no sound; it is magnitudes quieter than the quietest place on Earth, so that even if you were to shout you wouldn't be able to hear your own voice.

Imagine further that this world has been battered for billions of years, in a relentless shower of debris, so that its entire surface is scarred (there's a barren beauty about it). Traces of these collisions remain fixed forever, though most of the celestial objects (traveling at speeds up to 25 kilometers per second) have been vaporized upon impact. What's survived has shattered into fragments that suggest almost inconceivable violence. (Consider, if it helps, that science tells us the energy of impact—or kinetic energy [KE]—depends on both the mass of the object [M] and the impact velocity

[V] such that KE = 1/2 MV².) So even the tiniest projectile, should you find yourself in its path, would pose a considerable problem.

Additionally, suppose that the temperature of this world ranges from -240 to +240°F depending on its position relative to the Sun.

Question: If offered a seat in a spaceship, would you want to visit this alien world?

And if you were a scientist who had made its study your life's work, would you want to take the trip, as one did, even if you could never come back?

◦ ◦ ◦

Strictly speaking, this isn't a hypothetical any longer, since mankind has now traveled to the moon six times. But for millennia it had seemed impossibly distant. There were always *lunatics*, of course, who looked up at the sky in wonder, some of them at Stonehenge and some at places like Mount Palomar, employing the latest technologies to get the best possible views.

In 1856 William Crookes, Esq., reports to the Royal Society on a new field he calls *lunar photography.* Crookes is a self-described "scientific man" who sees in the camera a useful tool for advancing celestial studies. One will no longer have to make do with crude drawings or descriptions of the moon; it can be captured as it really is, without reliance on man's enfeebled faculties.

He's convinced the Society to underwrite his efforts, and his *good friend Mr. Hartnup* has given him the run of the Liverpool Observatory. It's only a question of perfecting the process.

He works in a cold room (in fact, as he notes with a tinge of complaint, it's well below freezing). Once he's exposed the negatives he plunges them into the special solution he's prepared: crystallized nitrate of silver, *perfectly pure and neutral*, mixed with iodide of cadmium and collodion in water. It works, but he wants the Society to know that it wasn't easy; *although this process seems very simple, it is impossible to estimate the difficulties I had to overcome.*

He's pleased with his negatives. The images they produce are by all reports beautiful. They would have been even better if technology could keep pace; he laments the lack of a paper fine enough to render the detail he's obtained. The slightly fuzzy prints he makes seem marvels of the modern age, but he knows better. For the moment, the sharpest images of the moon ever captured must remain in his mind's eye.

He won't live to see the bittersweet culmination of such efforts.

In 1964 the unmanned Ranger 7 spacecraft will take more than four thousand photos of the moon until just .12 seconds before it crashes into the surface. It's moving at 5,850 miles per hour. It will be obliterated, unavoidably, a state-of-the-art instrument sacrificing itself for science. But up to that very last moment it will send back pictures of stunning clarity, thousands of times sharper than any previous image of the moon.

What's left of Ranger will come to rest, fittingly,

in an area known as Mare Cognitum, the "Sea of Knowledge."

October 4, 1957: *Sputnik*, the world's first artificial orbiter, hurtles at 18,000 miles per hour around the planet. It's just a twenty-three-inch aluminum sphere, a shiny bauble designed to display itself as it catches the light of the sun. The Russians want the world to watch. And if you can't see it, you'll certainly hear it as the steady *beep beep beep* is played on radios around the world. It's a kind of mechanical heartbeat; it makes the *satellite* (a new entry in the vocabulary of popular science) seem to be alive. And that makes people wonder. If they can put a sphere of metal into space, how long before they send up a missile? Or even a man?

Everyone's been caught off-guard.

Some, of course, doubt that such capabilities are possible in 1957. *I am of the opinion*—Sir Harold Spencer Jones declares in *New Scientist*—*that generations will pass before man ever lands on the moon*. He isn't worried by the portent of what President Eisenhower himself dismisses as *a little ball*.

But others think that's a feint, it's Eisenhower trying

to calm the troops, he must understand that *Sputnik* changes the whole equation.

Americans know almost nothing about *outer space.* How could they? They haven't taken math or science courses, no one's told them that the future will be off-world, that their descendants will use tricorders instead of rotary phones. They don't yet know that someday Earth itself may be pulled into the Sun, and that mankind may need a new home.

It's a lot to think about, especially if you find the idea of other worlds unsettling. It's one thing to ride on an airplane—at least you're still close to the ground—but even that form of travel isn't yet routine.

A new curriculum will be developed for schoolchildren (their supple minds feed on the fantastic) but too many of their elders will find themselves closed off from the excitement. They possess only a cramped understanding of what's possible. And you can't look to the future if you can't even conceive it. Some scientists attempt to reach those who wish to understand, beginning with the basics: *the force we call gravity results from the distortion of something we call space-time—which is not simple to explain.*

It's taxing for experts to speak in the language known as *nontechnical*.

But some of them are still willing to try.

In 1958 the Legislative Reference Service sketches some early unknowns about the challenges of space travel. It's a preemptive inventory of unresolved issues, veiled in the bland discourse of the bureaucratic.

Question: *What information does the Congress require for its own use?*

Translation: How can legislators fund something they don't understand, especially if they're loath to admit it? It's one thing to suggest that the public requires remedial instruction, but you can't call a congressman an idiot and expect to get an appropriation.

Question: *On a space flight what are the problems, individual and social, of a psychological nature?*

Translation: How reliable will humans be in a radically new environment? We have some idea of how a machine behaves in space, but who knows what will happen to people? They'll experience tedium, stress, and most of them (if they're normal) even fear. What kind of person can handle it? After all, we're talking a massive investment of time and money.

Question: *What encouragement should be given to amateur rocket enthusiasts?*

Translation: We're in trouble—maybe some kid in a garage has an idea. Who knows how much time we have left to catch up?

In 1959 the Select Committee on Astronautics and Space Exploration, charged to create an official report for Congress, publishes a 241-page *Space Handbook*. It begins with a teaser: *perhaps astronautics will show man that he is not alone in the universe.* That's enough to entice even a skeptic to keep reading.

The report proceeds with a simple illustration of the solar system, as if to remind congressmen long out of school just what's out there. It's like studying a map of your district—here's Earth, there's Mars, just across the way, and there's Saturn, represented by its rings. At the center of everything, of course, is Mr. Sun, so big he can't even be pictured fully on the page.

At that point the reader, lulled, is confronted with more challenging information: *the Sun is a "main-*

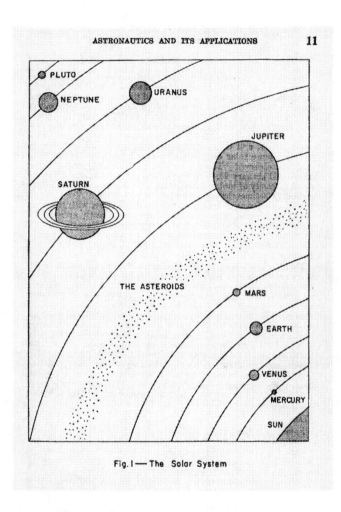

Fig. I—The Solar System

sequence" star of spectral type G-zero with a surface tem-
perature of about 11,000 °F. It helps to know the neighbor-
hood. Which we'll never leave, if we don't also know

that the *escape velocity*—the speed required to escape from Earth's gravitational pull—is about 36,700 feet per second.

And then just a few more numbers, a quick dip into relativity theory. The whole prospect of long-term space travel, after all, depends on this foundational idea. Maybe the concept can be snuck into a little story:

Consider 2 men, A and B, of identical age, say 20 years old. Suppose B took a round trip to the vicinity of a nearby star at a speed very near that of light (about 186,000 miles per second). It would appear to A that the trip took, say, 45 years—he would be 65 years old when his friend returned. To B, however, the trip might appear to take about 10 years. He would be 30 when he returned.

Chances are you've just lost even the most engaged congressman. It will be much easier to focus instead on the idea of *a space race*, of beating the Russians.

Everyone understands that.

The Lunar and Planetary Exploration Colloquium, 1959. Finally, scientists can talk to each other. What a relief to make an intelligent remark without contorting one's thoughts to the lowest common denominator (itself an overly simple metaphor). Here, someone can declare that *rocks are extremely complex systems* without risking derision. In fact, he's likely to find the audience nodding in assent. They know that what looks like a gray lump to a layman is an intricately detailed case study to a geologist, revealing, potentially, the secrets of an entire planet. Invariably, the rock itself is more interesting than what's under it (an amusing observation that few laymen are likely to find funny).

Those assembled debate the scientific value of manned space missions. Some want to send unmanned probes to distant worlds. They see man as an expensive liability in space, a fussy mechanism that's difficult to maintain. Their vision puts the science first, but that's never going to fire the imagination of taxpayers in the 1960s. It won't be until some time in the early twenty-first century that personable little probes, with names like *Opportunity* and *Spirit,* dazzle the public as they scoot about the surface of a foreign world.

Better for now to accept the *astronaut* as a given and move on to other issues, such as the search for extraterrestrial life.

First off, the question of criteria: *What should we look for?* Most people in the late '50s would think of a bloblike alien that oozes, or an articulate stranger who speaks English. He/it would be moving around in a souped-up saucer. But neither one of these models is what the scientists will be searching for. They understand that *life as we know it is a manifestation of certain molecular configurations*—most likely a variant of a carbon compound. It might even be undetectable unless they build the right sensors, which depends on making the best guesses now.

We can't instrument for things that we can't predict.

Whatever it looks like, it could well be scarier than anything in a science fiction fantasy. One has to worry about bringing back an alien organism, inadvertently, something that might consume our oxygen, or infiltrate our food supply, or even, conceivably, vie with our species for planetary dominance.

It would be a mistake to characterize such an organism as unfriendly; it probably wouldn't have any particular attitude towards us at all. The old paradigms—the agitated conflicts of human societies—aren't likely to obtain in an impersonal universe. Emotion, like consciousness, may be a local phenomenon, confined to our little corner of space.

So a consensus is building for some kind of non-psychological microbe.

The scientists also talk of the need for a *space suit*, a casing to shield the astronaut from an environment that isn't made-to-order for man. He needs to be kept warm, and cool, and protected from wayward particles so minute they cannot be seen without a microscope. The tiniest leak would cause air from the suit to spill into the vacuum of space, and the bodily fluids of the astronaut would then begin to boil.

He wouldn't be coming home.

Other worries: On landing, the lunar module itself might sink into the moon's surface; the lunar soil cling- ing to spacesuits might spontaneously combust in the module's cabin; the human body might simply fail to function in the weightless conditions of space. We might find that the moon—close as it is to our home world —proves inhospitable in a thousand different ways.

But that's assuming we can get there.

The moon is 238,000 miles from the Earth, and hitting that target is *nontrivial*. Neither body will be station- ary, and the spacecraft itself will be moving at thou- sands of miles per hour. So you'll have three celestial bodies in motion, linked only by the mathematics.

It's far too complicated a course for a pilot to plot; it's got to be worked out by the guys on the ground. And that's just getting there. Reentry is daunting too, you don't want to overshoot Earth and spin off into deep space. The *reentry corridor,* as it's called, is only forty miles wide. Getting that right, as one astronaut conjectures, is *like trying to split a human hair with a razor blade thrown from twenty feet.* And that's probably an understatement.

You'd better be able to trust the equations.

All that admitted, the scientists can't wait to get started. They've got the concepts and the tools. It will just take some hundreds of thousands of man-hours to *work the problems,* as they like to put it.

In 1961 the experts give the president-elect a final report on the status of American prospects in space. Focus on the positive first: A spectacular discovery of alien life would be *one of the greatest achievements of all times.*

But there's a worst-case scenario as well. The death of an astronaut in a failed mission would certainly lead to a serious national embarrassment. And while the whole world watched.

Then again, the space program could lead to improved communications capabilities; everyone could enjoy a level of TV reception available now only in major urban areas. That's a feat that would unite the world, making it possible for later generations to bring figures like ET and Alf into their own homes.

A very large booster rocket will have to be developed, but von Braun and his boys have signed off on that part of the problem.

On balance, the commission recommends a full-scale round-trip to the moon. The timeline will be tight, but it can be done, and we're the ones who ought to do it.

Accordingly, on May 25, 1961, President Kennedy announces his decision to Congress: *I believe that this nation should commit itself to achieving the goal, before this decade is out, of landing a man on the moon and returning him safely to Earth.*

The years will fly by in a flash.

July 20, 1969: Over a million people have come to watch the liftoff of Apollo 11, first manned mission to the moon. Many have pitched tents out on the open land around the Florida launch site, the empty acreage now littered with station wagons and trailers and vans.

(Only after the launch will the astronauts themselves be able to see, through their small window, the crowd spread out for miles like a vast, animated organism.) Most everyone's in shorts and bathing suits. There's a picnic basket over here, a Thermos of weak coffee over there, and maybe something stronger in the cooler for later, after the astronauts have gotten safely off the ground.

The trip to the moon will take three days, no sense waiting till then to pop a cool one.

Lots of families, no hippies (the VW bus isn't psychedelic). It's a party, but it's also a privilege just to be here. Most of the five hundred million people who want to witness this moment will have to see it play out on TV. Of course, most of them won't tune in until later, when Neil Armstrong steps out onto the surface of another world.

But the path to this point won't be completely uneventful.

January 27, 1967: The three astronauts of Apollo 1 —Gus Grissom, Ed White, and Roger Chaffee—are deep into preparations for the flagship flight of the moon program. Apollo 1 is charged with conducting a shakedown of the new Apollo/Saturn system, from pre-launch to Earth orbit. The fifteen million separate components of the machines, together with teams of technicians on the ground, have to work flawlessly if the countdown to the moon is to proceed. Every system has to be double-checked, every procedure picked at and tweaked, even if they do have a deadline. No pilot wants to go up in a vehicle that hasn't been tested and tested again to the point of tedium. And no one on the ground wants to be responsible for a *nonnominal* event in flight.

Today they're working the plugs-out test.

The astronauts are fully suited-up, strapped to their couches and hooked to hoses. It's a little cramped, but capacious by comparison with the Mercury and Gemini spacecrafts.

They've watched *the bird* grow—monitored its construction at aerospace plants, posed alongside it for promotional pictures, imagined themselves floating in it far above the Earth.

They plan to launch for real in less than a month.

They don't yet know that liftoff will never occur, that today the capsule will kill them.

Years of planning will come to this, a fire lasting less than thirty lethal seconds.

An elaborate forensic reconstruction of the catastrophe will be conducted, involving hundreds of interviews with personnel on the launchpad. Five thousand pictures will be taken of the crippled command module as it's stripped down, piece by piece. For comparative purposes an identical spacecraft will be taken apart as well. The painstaking disassembly will take two months.

Finally, another fire will be ignited in a full-scale mock-up that duplicates the placement of *combustibles* in the doomed Apollo. They'll find that packing Velcro, nylon webbing, and polyurethane foam into the spacecraft wasn't a good idea.

For the record, eyewitnesses add their piecemeal impressions to the grim picture that's slowly developing.

I saw flame billowing from the Spacecraft . . . a tremendous flash . . . Just one big boom, one big blast.

The team leader recollects what he heard in the first frenzied moments. *Fire in the command module,* or *fire in the spacecraft,* he's not sure just how the crew worded it in that final transmission. *The term* fire *stuck out more than anything else.* His white coat was peppered with embers that burned holes, *like cigarettes.* His checklists were charred.

Just five minutes after the fire is first reported, the inner hatch—the last barrier to the crew—is pried open. The team leader reports that he cannot describe *the situation* inside the command module. It's improvised code. He's speaking over an open channel, and he doesn't want to be the one who tells the world there's nothing alive in there.

Later, for the review board, he's asked to be more specific.

It appeared that their suits were shredded. I could see bare skin.

Another eyewitness is also pressed for details.

There was nothing but what appeared to be a blanket of ashes across the crew couches.

The board's report eventually notes the *extensive fusion of suit material to melted nylon in the spacecraft.* They'll need to design a new spacesuit, with a material that won't feed the next fire they hope never to have.

The technicians, stricken, want investigators to understand that they couldn't help what happened. *Personnel used all fire bottles that could be found.* It was the gas and smoke that drove them back. *There was five of us out here, and we each took turns going in and going out, trying to do something.*

They attempt to help with the timeline. *I think I got scared again. Everywhere I looked there was flame.*

One of the men is asked to elaborate. It's really only the shock that he remembers.

I may be able to amplify the first thirty seconds a little bit more if you wish me to but it happened so fast, it's hard to explain.

The nonhuman data readout is more precise. It tells them that moments before the flare-up a 1.7-second dropout in signal from the C-band decoder and transmitter outputs occurred, as well as a brief dropout of the VHF-FM carrier, a fluctuation in rotation controller null outputs, and a fluctuation in the gas chromatograph signal.

That's Apollo 1, dutifully reporting its own demise.

In order to create a complete dataset of the accident they'll need to clarify the last five seconds of audio transmission from the astronauts, but the tape is garbled. They won't get much assistance from the experts at Bell Telephone.

The present state-of-the-art analysis of voice records is such that little if anything can be determined as to what was said if the recording is not sufficiently clear to be intelligible by listening alone.

In other words, science can't help them. They'll have to rely on the old-fashioned human ear.

Most listeners decide that the astronauts' final

words were either *Let's get out—Open 'er up,* or *Let's get out—We're burning up.* The audio analysis ends inconclusively.

Some people feel that the very end of this second transmission is a scream, or the start of one.

But that could be the human bias.

The specific cause of the fire will never be established, though *an electrical arc in the sector between −Y and +Z spacecraft axes* will be tapped as the likely suspect, abetted by a preponderance of pure oxygen. Dozens of other design issues that need to be addressed will also be identified.

The mission patch of Apollo 1 will be known as *the patch that never flew* until it's taken to the moon by the Apollo 11 crew aboard a thoroughly redesigned and much safer spacecraft.

◎ *Thought Experiment*

Suppose that you can't foretell the future, and have no way of knowing how it will work out. Suppose further that you have a reasonable tolerance for risk, but aren't reckless. You know how many things can go wrong in the most complicated mechanism ever engineered.

Question: If you had already lost three of your buddies, on the ground, would you still want to go into space?

July 20, 1969: According to the Apollo 11 Flight Plan, touchdown of the lunar module *Eagle* is scheduled to occur precisely at 102:47:11 EDT. (It's Sunday on Earth, but that designation doesn't make much sense in space.) The flight plan is sprinkled with helpful doodlelike drawings to keep the astronauts oriented; at this point, if all has gone as planned, the LEM will be resting on the moon's surface, having survived a succession of *GO / NO GO* decisions. (It will seem in

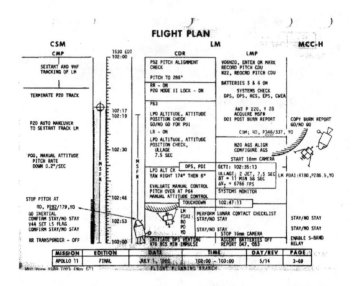

the drawing like it's sitting upside down, but that too is an Earthcentric perspective.) Once on the moon, the astronauts will perform a final Lunar Contact Checklist before Houston makes the crucial *STAY / NO STAY* call.

They're only on page 3-69 of a book that extends to 135 pages.

Man stepping out onto another world: It might be the most dramatic moment of the twentieth century (some say of all history), the moment Earth-bound creatures have imagined for millennia, and it will only be

possible if it's mapped out in micro-increments, with nothing left to chance or a faulty human memory.

It's a narrative made up mostly of numbers.

The flight plan was really very well written, Neil Armstrong himself will acknowledge. He might be its most avid reader. Some of its sequences have been stamped in neat block letters on his spacesuit sleeve, reminding him to UNSTOW BAG/SCOOP/HAMMER, UNPACK SRC. Houston wants near-zero probability of an unscheduled surprise.

They've picked the right man. The thirty-eight-year-old pilot from Wapakoneta, Ohio, is famously focused. He believes in the trinity of *training, planning, testing*— running as many "what-ifs" as possible until the direst simulation triggers a default response. There's protection in procedures. No sense hitting the panic button.

You've got your job to do and you just go ahead and do it.

But planning can't account for every contingency. He's at two hundred feet now, and the computers have him headed into the heart of a monster crater, surrounded by boulders, with less than thirty seconds of fuel left. In this landscape the lunar module will probably tip over on landing, like a skittish spider, or find itself settled deep in a hole it can't climb out of.

Either scenario and they're stranded on the moon.

Gonna be right over that crater, he tells Houston. It's his first *uh-oh* moment of the mission. He doesn't say much else, but the data speaks for him. His heart rate, which has averaged just 71 beats a minute, jumps to 100 during the descent. *On a scale of 1 to 10*, he'll explain later, *that was a 13; there were just a thousand things to worry about.*

But that's leisurely after-the-fact analysis; right now, he has to fly the LEM to smoother ground. They're running on fumes. He takes manual control of the spacecraft and does what he's always done in a thousand tight situations on Earth.

Houston, Tranquility Base here. The Eagle has landed.

In a celebrity-hungry age he becomes the most famous man in the world. But he's wary of attention, he's *spring-loaded to the suspicion position.* He thinks other people are *a third-rank category* of things to talk about, and seems even less interested in himself. Under duress, over time, he'll reveal a few begrudging personal details. He read the magazines *Flight* and *Model Airplane News* as a boy. Lindbergh was a great pilot. *The Right Stuff* was entertaining, but *bears no resemblance whatever to what was actually going on.* (He's too polite to use stronger language.) Stanley Kubrick is the only guy who got space right, in *2001.* That's a movie he could see more than once.

Journalists and historians will ply him for the rest of his life with questions that beg for a third-rank response, some admission of emotion; even a few fellow astronauts will express unease. *Neil never meets anyone halfway.* It's almost inhuman. But he'll dart around the possibility of disclosure with a skill undiminished over decades, like an early-issue cyborg that's still in top shape. He'll keep the focus where he thinks it belongs, never saying much more than what he told the world on TV, in 1969, as Apollo 11 made its way home: *the responsibility for this flight lies first with history and with the giants of science who have preceded this effort.*

He knows it was never about him, it was always about something much bigger.

The scientists themselves are most interested in the moon rocks Apollo 11 returns.

They've spent years devising protocols for *lunar cura-tion*, a new process of inspecting and preserving mate-rials of non-Earth origin. The Sample Return Container, or SRC, is really a big rock box, nothing complicated about that, but it's the first crucial link in the chain. It will be sealed in a large plastic bag upon landing and deposited in a special *Planetary Materials Laboratory*, engineered to the most stringent specs possible, with vaults designed to *withstand tornadic winds even though the rest of the building may be demolished*. A building can be rebuilt; the material it shelters is irreplaceable.

Protocol dictates that two people witness any trans-fer of lunar samples within the facility, but they can't get near the specimens unless they wear coveralls, caps, gloves, and booties. The human body is a mobile petri dish, host to a welter of contaminants easily transferred to whatever it touches (another argument for robotics). But the filtered air showers and vacuum locks should help.

Scientists have forty-eight pounds of material to work with, less than they'd like, given that one thou-sand researchers are waiting for their own chips and slivers to examine. And political egos have to be fac-tored into the equation. VIPs are already bidding for their own tokens—extraterrestrial tchotchkes destined for display.

Eventually, everyone else gets a chance to take a look at history up close.

Thousands of expectant citizens queue up outside the Smithsonian when the moon rocks are finally placed on public view. They're happy to stand for hours, the wait will be worth it. They expect something dramatic, otherworldly, something that glitters or sends out signals or maybe even moves, if imperceptibly. Not a drab chunk of gray they can find in their own backyards.

It's unbelievable, *it looks like any rock.*

In their own way the scientists are disappointed as well. They test the lunar samples for every physical and chemical property possible, with the most sophisticated instrumentation available on Earth. The results are indisputable—the samples contain no trace of fossil, germ, or biogenic material whatsoever.

In other words, no life.

It's a blow, but no reasonable person can interpret the data differently. The moon is a dead world, like the rest of the known universe, with the lonely exception of Earth.

None of the six remaining manned flights to the

moon will uncover any further evidence to the contrary. It will be left to the visitors themselves to leave traces of life on the moon's surface, as abandoned lunar rovers, spacecraft, and experiments begin to collect haphazardly, the detritus of first-phase exploration. The space junk is expected to remain undisturbed for eons.

One item will be deposited on the moon's surface with greater care, though it's now considered unlikely that any other living species will ever see it. It's labeled, just in case.

THIS IS THE FAMILY OF ASTRONAUT DUKE FROM PLANET EARTH. LANDED ON MOON, APRIL 1972.

10

◎ *Thought Experiment*

Imagine you're in a pre-K classroom. A play session has just started. Most of the children have gathered together in groups, but one child is off to the side, alone, concentrating on a large construction built out of blocks. It's irrelevant, in this context, that you didn't like the lunch you just ate and wish you'd gotten something else, or that childhood will prove your least favorite period of life.

It also isn't material that you might prefer to play with Tinkertoys or Legos.

Question: Are you in the larger group of children, or are you the one sitting off by yourself?

Are you male or female?

Is your construction haphazard, or arranged in symmetrical patterns, with spandrels and setbacks?

Are you happy that your block world is uninhabited, or do you wish you had little people to move around?

Do you find yourself sneaking looks at the crayons and paper across the room?

Your answers may determine which side of the cultural divide you eventually end up on.

Research suggests that play can spur imagination and boost mental development. In fact, it's one way to identify scientific aptitude. The great scientists of the past lost themselves as children in spools, sticks, pieces of wood and wire; the most promising scientists of the future are almost surely playing somewhere, with something else, right now.

It's life's first lab.

And those who take such play seriously might be able to keep at it their entire lives.

But studies also indicate a gender difference in attitudes towards play. If you're a little boy, you like to build things or move them around; if you're a little girl, you like to play with things you can cuddle or rock. In the one case, you're *systematizing*; in the other, *empathizing*. One of you is more comfortable with mechanisms, the other with feelings and people.

But there's disagreement over the data regarding

gender distinctions. And evidence is building that such thinking no longer serves us as a species.

As humankind evolves, one can imagine a world in which the old divisions blur. We may all someday care about machines, and see ourselves as systems. We'll have to weather some post-Cartesian confusion, but at some point a new being may emerge, designed to the specs necessary for survival.

Someone must already be working on that, somewhere.

August 11, 2003: The first successful transatlantic flight of a toy, a triumphant solo by an eleven-pound model airplane cruising on just one gallon of gas. The

Spirit of Butts Farm, made of balsa wood and Mylar, has flown from Newfoundland to Ireland in thirty-eight hours and fifty-two minutes on autopilot. Until the last moment no one could be sure that the diminutive model would make it. *A great cheer went up when we saw it*, one of those awaiting it on the ground recalled. *The last hours of the flight were "white-knuckle" for all of us.* At landing, the plane has just 1.8 ounces of fuel left, a margin so slender that it's a miracle. Or maybe just a marvel of engineering.

The flight sets new world records for model airplane aviation.

It's hard not to admire the plucky little plane, flying by itself for almost two thousand miles, a bright speck of red in the sky. It has to overcome myriad challenges, not least the sheer mechanical monotony of an extended flight. But grit isn't all the *Spirit* needs. It helps that the autopilot (a miniaturized mechanism five years in the making) consists of microchips attached to a GPS receiver, an altitude sensor, and a piezoelectric-rate gyro that monitors angular velocity. Further modifications to the fuselage, engine, propeller, and fuel have resulted in a trim craft that can cruise at 49 miles per hour and outfly every other model in its class. You'd need that kind of innovation to pull off what experts consider *the most ambitious project in model aviation history*.

The idea is the brainchild of retired engineer Maynard Hill. He's the ghost in the machine, its seventy-seven-year-old guiding spirit who never grew up. The infirmities of age have ravaged his body—he's nearly blind and can hardly hear—but the plane he designed suffers no such limitations.

In a sense, they've both taken the trip.

October 2004: Software engineer John Pultorak emerges from his home workshop after immersing himself in the most ambitious solo project he's ever undertaken.

He's spent four years, in his spare time, building a replica of the forty-year-old Apollo Guidance Computer, or AGC, the system that took men to the moon. The original is an antique now; a modern cell phone or ATM has magnitudes more computing capability. In fact, it's so rudimentary that Pultorak has posted instructions on his website for like-minded hobbyists: *Apollo Guidance Computer—How to build one in your*

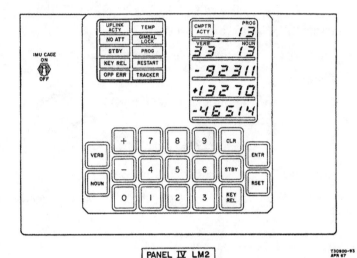

PANEL IV LM2

T30900-93
APR 67

basement. He thinks it's instructive to learn a system from top to bottom, and vintage computers aren't impossibly complex, like their modern counterparts. *You can (if you like) actually understand the entire computer, from hardware to software.*

He knows how rare it is, even for a pro, to understand something fully in science.

With a little additional online investigation, hobbyists can also find the original AGC design documents from MIT that NASA has now declassified. It's a trip back in time. The typewritten reports and hand-drawn schematics were drafted by the best minds of the day, working what a later age will term 24/7, attempting the seemingly impossible.

Conventional computers of the early 1960s fill entire rooms; any computer installed in a spacecraft must be radically miniaturized. They'll use the new concept of *microelectronics* to design tiny chips and integrated circuits.

The AGC has a one-megahertz processor and a single kilobyte of RAM. (Translation: It can get to the moon, but it wouldn't be able to handle MP3s or frequent postings on Facebook.) It's programmed in a new code known as *basic machine language*, with inputs intended to mimic the syntax of spoken language—verb, noun, and number keys, and an Enter button to initiate a sequence. The astronauts will spend months training to enter strings of instructions accurately. *Verb 33, Noun 13; ⁻92311, +13270, ⁻46514; ENTR.* The user interface has proven to be a hotly debated area of design; there's inevitable tension between the capabilities of the machine and those of its operators. The designers add flashing lights for moments when the AGC itself needs *to attract the astronaut's attention.*

All things considered, it's better to design around the human element whenever possible.

The computer has self-diagnostic programming intended to detect 98 percent of any faults that may occur. Should an in-flight repair be required, the engineers at MIT add this proviso: *It is assumed that step-by-step instructions will be available for the astronauts.*

They run a series of punishing tests to ensure the computer can survive the extremes of space—flight shock, thermal vacuum, Earth-landing shock. It is *overstressed* in subroutines intended to push it to failure. When they're finished, the fifty-pound AGC, primitive as it is, will do its job with distinction as it travels alongside the astronauts into space.

Forty years on, the instrument is still formidable. The Pultoraks prepare to fire up their own AGC for the first time.

We all went down and watched it blink. It was awesome.

They get offers to buy it, but that's unthinkable. It isn't going anywhere; it's a member of the family now.

May 11, 1997: Garry Kasparov, world chess champion for the past twelve years, arguably the best human

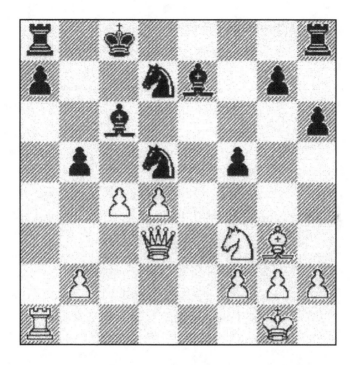

player in history, sits for the sixth and final game in a match against the IBM supercomputer Deep Blue. He'll be playing black. This isn't the first time a computer has tackled chess; the IBM 704, Blue's distant ancestor, pioneered such play in 1958. But Blue is a million times more powerful than the 704. Scientists have worked relentlessly to make it so, tweaking its circuits, amplifying its memory, striving to imbue it with something that approaches understanding. The complexities of chess provide the perfect means to explore higher-order cognitive functions in artificial intelligence. They consider Kasparov a member of their scientific team, but they badly want to beat him.

The score is tied at two-and-a-half games apiece.

The champion knows what's at stake. For him the match is *species-defining*; it poses questions that go to the very heart of what it is to be human. What does it mean to think? Is the human brain unique, an exquisitely complicated network of nonreplicable neurotransmitters, or can a machine mimic its functions, someday even surpass them? Kasparov isn't just battling for himself, a victory by Deep Blue would be *a frightening milestone in the history of mankind.*

The scientists know that Blue can beat 99.999 percent of all human players. It's the rare specimens like Kasparov that interest them.

Bobby Fischer was once called *the greatest chess machine mankind ever invented,* before his human demons destroyed him. Kasparov is the real man-machine, steady rather than mercurial. His encyclopedic chess memory has been built up over a lifetime of absorption in the game. A player at his level intuitively accesses thousands of *little chunks of chess knowledge*, discerning patterns where lesser players simply see pieces. There are billions of potential combinations on the board (or, more exactly, 10^{40}). Kasparov can sometimes calculate two moves per second. That's the kind of blinding speed that causes his opponents to despair.

And a major reason why he beat an earlier version of Blue in 1996.

But this new Blue can review two hundred million positions in that same second, flying through its own database of seven hundred thousand grand master games. It's called brute force calculation, more brawn than brain. Blue's cutting-edge search algorithms make it an idiot savant of sorts, fearsomely effective, but still *less intelligent than even the stupidest human.* It knows nothing about insight or strategy.

In science, it's one step at a time.

Kasparov is concerned. He doesn't know what they've done to Blue this time around. Like any canny player, he tries to psych out the other side before they begin. He insists that the match rules forbid *any distraction that may reasonably be regarded as disturbing.* He knows he's at a disadvantage—Blue doesn't care about disturbances, *it never tires, it never gives you a break.* He's got to make sure that his concentration holds up, that his body doesn't subvert him, that his mind is able to bore deep into itself. That's where the match will be won.

Chess experts are asked to provide play-by-play commentary on game six as it unfolds. One of them has a gut feeling that Kasparov will prevail. The others aren't making any bets.

The match begins. Soon it's apparent that something isn't right. He's off his game.

Kasparov has been rattled, Kasparov is shaking his head as if something disastrous has happened.

He knows before they do. It takes them a few moments to catch up.

Yes, he has blundered, now the king cannot move. Deep Blue has an ideal attacking formation.

Just one moment of weakness.

Checkmate.

It will be the shortest loss of Kasparov's career. Like everyone else, he's shocked by how well Deep Blue has played.

o o o

Soon after the match ends Kasparov accuses the computer of cheating; he insists that some of its moves were too unusual, too intuitive, for a machine. It had to have human help. But he can't prove it, he has only a hunch, a gut feeling, and that's not enough.

He asks for a rematch, he even proposes to put his title in play, but IBM isn't interested. Its scientists want to move on.

Kasparov eventually capitulates and extends his highest accolade to the mechanism that's beaten him. *Deep Blue,* he admits, *is a grand master.*

And Blue? It's already on the scrap heap. It will live on, in some fashion, in the sophisticated chess software of the future, available to anyone with a home computer.

11

MOST PEOPLE SHY away from chess, but why are they morbidly afraid of math?

Some suffer from *innumeracy*, a simple ignorance of mathematics, while others are *dyscalculic*, a more serious condition linked to a deficiency in the left parietal lobe. Nonmathematicians have their own word for it, *dumpkopf,* a descriptive term with such little sting, in this context, that otherwise prideful people freely apply it to themselves.

Can't balance a checkbook? Don't understand decimal points? Can't do differential equations? No shame, nobody else can either. And that's just the low-level math one was supposed to have learned in school; the higher-level reasoning that experts consider real mathematics is out of almost every mortal's reach. According to informed opinion, there's just one memorable mathematician for every four million people. The rest of us, even those adept at formulas and algorithms, are *playing Chopsticks* by comparison.

But even the memorable mathematicians can't work at that altitude forever.

G.H. Hardy, dour author of *A Mathematician's Apology*, famously berates himself as his own mathematical powers wane. He's sixty, long past the age of youthful glories, his once nimble mind gone to mush. But he finds some consolation in his decline; by his accounting, *most people can do nothing at all well*, so even the faint memory of important work is sustaining.

Hardy's assessment is harsh, it's the kind of predestination argument that discourages people from making the effort. But chances are most of those millions of potential mediocrities have no inkling just how skilled they actually are. As toddlers, they begin by counting, painfully, one finger for one number, one number for one thing; over time, in a miracle of mental development, even the average human makes a leap, the sort that separates him from other species. Numbers become malleable, they can stand for anything; throw in a zero, and one doesn't need fingers anymore. It's the beginning of a feel for abstraction, the crucial prerequisite of all scientific progress.

Eventually, the most confident proceed to advanced puzzles like this: If two typists can type two pages in two minutes, how many typists will it take to type eighteen pages in six minutes?

It's a question taken from a Mensa test. Solve it and you'll become a member of the class of self-identified smart people, the kind who would never join the much larger *dumpkopf* club.

As it happens, the history of mathematics is studded with unsolved problems, famously frustrating theorems that have inspired, over centuries, obsessive quests for a proof. The stakes couldn't be higher; anyone who puts such a problem away is guaranteed mathematical immortality. And even a misbegotten attempt might advance the state of mathematics. But that's a collateral consideration. The real reward is the quest itself.

It's like Ahab chasing an equation, sometimes at the cost of an entire career.

◦ ◦ ◦

Squaring the circle is a deceptively complex problem first posed by Greek mathematicians in the fifth century BC. It seems simple enough: *using a straightedge and ruler, in a finite number of steps, construct a square that exactly equals the area of a given circle.* Anaxagoras, Oenopides, Antiphon, Hippocrates, and Hippias will take up the challenge and fail; over the next several centuries Indian, Chinese, and Arab mathematicians will take their own crack at it; the problem will frustrate Descartes and Newton as well. None of them will solve it.

Many will provide approximate solutions, but in classical mathematics *almost* is not good enough.

Eventually, in 1882, it will be proven beyond dispute that the problem is impossible; they'll know about transcendental numbers by then.

In the nonscientific community, the phrase *squaring the circle* will become a byword, like *tilting at windmills*, for any effort that isn't expected to end well.

Those who love the field labor on, particularly the seventeenth-century French mathematician Pierre de Fermat. He's diligently annotating a copy of *Arithmetica* by Diophantus of Alexandria (circa AD 250). He's come upon a problem that confounded the Greek, but doesn't seem daunting to him. He makes a casual note to himself that will bedevil ambitious mathematicians for the next 350 years.

I have discovered a truly remarkable proof but this margin is too small to contain it.

It's the most famous teaser in mathematics, the solution to an ancient problem apparently alighting, for a moment, in the ephemeral medium of one man's mind. If the answer appeared to him, maybe it will be visited upon someone else.

Thousands of proofs will be proposed over the ensuing three centuries, each reviewed by experts and discarded. Some will attempt to reverse engineer Fermat's thinking, as it were, but that line of attack is unproductive. One can't work back from an enigma to a mystery and expect to arrive at an answer.

In 1993 the British mathematician Andrew Wiles will roil the scientific world by presenting a proof of the theorem that seems unusually promising. But nobody's risking a celebration just yet. Teams of experts will pore over the proof, line by line. The process is expected to take a couple of years. In the meantime, the editor of a math book with the misfortune of going to press in 1994 will acknowledge the development, if ever so cautiously.

If his proof is found to contain an error, it will join thousands of other might-have-beens. Stay tuned.

Wiles's proof will survive the scrutiny, and Fermat's last theorem will finally be taken off the table.

In the early nineteenth century another cryptic message is left to posterity by another brilliant mathematician. He's a twenty-year-old who's already done landmark work in the theory of polynomials, but he's got a duel the next day and worries that his ideas will be lost. He may be impulsive in love, but he's more deliberate about his work. That night he writes a letter to a friend.

I have made some new discoveries in analysis. There are a few things left to be completed in this proof. I have not the time.

In two days, given the exigencies of abdominal wounds, he will die, but the mathematical world will build upon his insights, instituting a fruitful new realm of thought known as Galois theory.

Meanwhile, in the field of math pedagogy, new

studies reveal that "human interest" math problems— the ones that tell stories, with trains, ships, and people passing each other on the slightest pretext—may inadvertently prevent students from understanding mathematics. Such narrative-driven formulations cause them to focus on the problem itself, and not on the concepts that underlie it.

It's like taking a trip, but never understanding the nature of travel.

12

◎ *Thought Experiment*

Imagine you've got an opportunity to examine the brain of a gifted scientist. Would it be bigger than an ordinary brain, more imposing in some manner? Could one isolate a special "science section" in it, something that lit up, as it were, whenever a scientific thought arose in one of its regions?

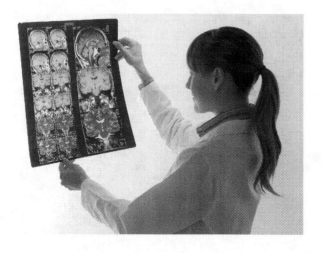

As it happens, Einstein's brain was preserved upon his death for just such research. Thanks to the work of Paterniti, Levy, and Abraham, we know something about the circumstances of the brain's disposition.

Science requires that we separate hard data from the merely anecdotal. So it isn't relevant, for our purposes, that the pathologist performing the autopsy removed the brain without permission, or that he kept it in his basement for decades, or that it was stored in a Mason jar, or a Tupperware container, or possibly both. It's also not pertinent that the technician, having lost his job, moved to the Midwest and befriended William Burroughs, or that the brain went with him.

Assume that the brain weighs 1,230 grams, slightly less than expected, but is otherwise unremarkable, save for an area associated with mathematical ability that is significantly larger than normal.

Question: Given his inherent anatomical advantage, how quickly would Einstein have found the missing term in the following series:

$$1 \quad 8 \quad 27 \quad ? \quad 125 \quad 216$$

Note: It's okay to take your time, research often proceeds at a snail's pace.

○ ○ ○

Truth be told, the brain of the average person is as likely to resemble that of a rat as an Einstein. Studies have revealed that rodent and human brains share a remarkably similar architecture. Electrodes don't lie. Once we accept the affinity, there's much to be learned about our own minds from such creatures.

Scientists know that rats running a maze, in their sleep, will replay the routes again and again, as if trying to get it right. Or trying to hold on, desperately, to some furtive bit of information that might later prove of use.

Bottom line: Their brains won't let them alone.

It's easy to extrapolate from that point. Sleep might actually be a stratagem, a pretext for some kind of demanding data processing. If that's so, then fatigue could be nature's way of fooling us. If we really knew how much work we'd have to do at night, we might choose not to sleep at all.

Research has also established that rats, like humans, dream in images, making their own little movies from the scraps of experience they've accumulated while awake.

It's the brain entertaining itself. Or maybe that's work as well, to an end that isn't yet clear.

Bottom line: The average brain is as busy as a brilliant one, but Nature, recognizing our tendency to indolence, would rather we didn't know that.

In memoriam: Alex the Parrot, *Psittacus erithacus*, 1976–2007. Pioneer of avian studies, exemplar of selfless devotion to science. And one smart bird.

He didn't work alone. Science is notable for great teams—Watson and Crick, oxygen and hydrogen, adenine and thymine, and now Alex and Dr. Irene Pepperberg. Together, over thirty years, they made groundbreaking discoveries in the study of *avian intelligence*.

And one of them was working with a brain the size of a walnut.

Alex (short for Avian Learning Experiment) came from humble, even unpropitious beginnings, having been picked up in a pet shop on impulse. Pepperberg had a theory. Other animal researchers had focused almost exclusively on dolphins and chimps, the usual suspects, but she was intrigued by parrots. In the wild, they'd been found to exhibit unusual inventiveness; perhaps, in the lab, they could offer a new means to

explore cognitive development. They were already noted for their language skills. It wouldn't be a case of mere mimicry; her work, if successful, would prove that parrots exhibited *intellectual acuity*.

In other words, "birdbrain" wouldn't be an insult anymore.

۵ ۵ ۵

The gatekeepers of science weren't initially open to the idea. Pepperberg found that no one wanted to fund her work. *People said birds were stupid.* Skeptics were resistant to the very possibility that an animal might possess something resembling awareness, something beyond instinct. The idea was preposterous.

But she moved ahead in spite of such discouragement.

Soon enough, Pepperberg and her assistants realized that Alex was unusual, perhaps even a prodigy. As the work progressed he learned numbers, colors, and shapes; he could identify scores of objects and knew 150 words. He made up new words on his own, like a parrot-poet. He seemed to grasp concepts. And he wasn't afraid to assert himself. He could be argumentative, even cranky. His trainers were impressed by his shifting moods.

On some occasions he just didn't feel like working.

No! I'm gonna go.

On the days he'd really had it, he would turn away from the trainer, or adamantly refuse a nut treat if he'd specifically asked for a banana.

That was a sure sign of intelligence. Any creature

could feel the primitive emotion of anger, but only a more complex being could express exasperation.

He also learned to make amends when he sensed he'd gone too far.

I'm sorry, come here, I love you.

That took intelligence as well.

One memorable day Alex saw himself in a mirror. Though it's impossible to say for sure, he seemed to be almost thoughtful.

What color?

After some hesitation his trainer responded.

Gray. You are a gray parrot.

He looked for other gray things after that.

Pepperberg never wished to overstate Alex's abilities, to exceed the evidence. He spoke, but *I don't believe years from now you could interview him.* She had scrupulously controlled over the years for experimental pitfalls—operant conditioning, unconscious cues from trainers, anything that could skew results. But after thirty years of testing, she could hardly shrink from what the accumulated data was telling her.

Alex knows what he's talking about. Alex can think.

⊙ ⊙ ⊙

Pepperberg and Alex exchanged good-byes, as usual, as she left the lab on that last night. He was terse, but no more so than normal.

You be good. I love you.

When she returned to the lab the next morning he was gone, peacefully, the familiar voice stilled forever.

She has other parrots to work with, young birds who show some promise, but she knows there will never be another Alex.

Such partners are rare enough in life, let alone in science.

13

○ *Thought Experiment*

Imagine you're living in the sixteenth century, somewhere in Europe, and you believe that the Earth you're standing on is motionless, the stable center of the universe. You know this for a certainty because you've always been told it's so, and anyone can see it's the sun that moves across the sky. (It isn't relevant, in this instance, whether you're a prince or a peasant, educated or illiterate.) Imagine further that you have a high regard for your own place in the Great Chain of Being, nearer the angels than the beasts. Everything has a place and a purpose, even if it's sometimes difficult to divine.

Now imagine that one day you hear word of an unsettling rumor. An obscure astronomer has advanced a new idea—he says the Earth moves around the Sun, the true center of the universe. He's got drawings, and mathematical proofs to accompany them, but he doesn't expect you to understand. He speaks as one astronomer to another; if you haven't been trained in the new science, you'll have to take his findings on faith. You won't hear his ideas praised in sermons. (In fact, his book will be banned by the Church for two hundred years.) You'll have to decide for yourself which version of events you believe.

Assume, for the moment, that you accept the new view.

Question: If the Earth isn't at the center of everything, where does that leave you?

Now imagine you've traveled through time to the twentieth century. (It isn't relevant, in this exercise, whether you've made the trip in a time machine, or via astral projection, or some other yet-to-be-devised method. It also isn't relevant that Europe is no longer the cultural or intellectual capital of the world.)

You have long since recovered from the Copernican shock you suffered in the past. You once again assume, almost casually, that humans holds a central place in the cosmos, though it may not be divinely ordained anymore; no likely competitor has come along, and it seems that the world you know has been tailored to the human scale. You can live with your planetary subordination to the Sun; it is, after all, special, even considered worthy in some cultures of worship.

All of a sudden, a scientist—perhaps an astrophysicist

—*proposes another new idea. He or she says the Sun isn't special after all, it's just average, one of two or three hundred billion such stars in your galaxy alone, which is, in turn, just one of hundreds of billions of galaxies in the universe itself. And each of these galaxies contains untold numbers of planets, which may or may not be similar to your own. In other words, you're bobbing on a tiny speck of stuff in an almost unlimited ocean, more a piece of driftwood than a result of deliberate design.*

Question: Assuming you accept this vision of a radically reconfigured universe, do you take comfort in your isolation, or would you like to have some company?

August 15, 1977: Jerry Ehman, astronomer, experiences a eureka moment while monitoring radio

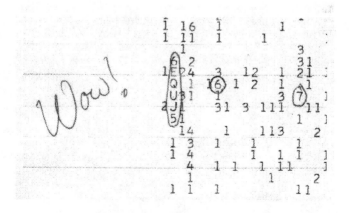

telescope readouts from space. He's working on a Search for Extraterrestrial Life project at Ohio State University. *Big Ear*, a large radio dish, has picked up an anomalous signal from the swath of sky it's been sweeping, something with all the marks of a non-Earthly transmission—it's unusually powerful, has a high signal-to-noise ratio, and lasts seventy-two seconds, conforming exactly to the expected parameters of an intentional interstellar message.

In other words, Wow!

Scientists have known for years that radio transmissions offer the only feasible means of communicating over vast interstellar distances. Even if one masters the technical challenges of propelling man beyond the moon (and it was hard enough to get there) the human lifespan is completely out of sync with the expected duration of future spaceflights. Just getting to Mars and back will take years of round-trip travel, though it's essentially next door, and without something like *warp drive* (a TV fantasy, sad to say) no human will live long enough to get much farther.

Radio waves, however, zip across space at the speed of light, in the blink of a cosmic eye.

And a Wow! signal is what everyone's been waiting for.

Lunch, sometime in 1950: Enrico Fermi, world-class physicist, is sharing a sandwich with some colleagues. They've been tossing around the implications of the age and size of the universe, loosely figuring the probability of alien life. Most of them are intrigued; they agree it's a significantly non-negative number.

But Fermi, the contrarian in the group (every scientific cohort has one), expertly upends the conversation with a single well-aimed question.

So where is everybody?

It's shorthand for not so fast, there's no credible evidence of alien visitations or communications, we have no actual data to work with. And if the universe is as big and as old as we think, they should have arrived already, or at least sent us a postcard from afar.

Fermi's offhand formulation becomes the Fermi paradox, a way of thinking about alien life that recasts the scientific debate. Skeptics like him are open to persuasion, but they won't be swayed by anecdotal references to Roswell or sworn sightings of oddities in rural areas.

For them, it's back to basics: Prove it, and then prove it again.

National Radio Astronomy Observatory, West Virginia, 1961: Dr. Frank Drake is scribbling on a blackboard, outlining an agenda for the First Annual SETI Conference. The search for extraterrestrial life has continued, unabated, in spite of a discouraging lack of cooperation from the extraterrestrials themselves. Now a group of stubborn scientists has decided to double-down, in effect, founding an organization to formalize their gut feelings.

$$N = R^* \, f_p \, n_e \, f_l \, f_i \, f_c \, L$$

What Drake puts on the blackboard that day will make him the point man in the SETI community for years to come. It's the Drake equation, a formula for estimating the likelihood of intelligent life in the vastness of space. He's come up with a string of variables that establishes specific parameters for such a search. Plug in some numbers—how many stars might have planets, how many of those planets might be like Earth, how many of those Earth-like planets may develop life, what fraction of those in turn may develop technology (a likely indication of intelligence), how long those civilizations may last, more or less—and you might arrive at a rough estimate of likely targets.

It's guesswork, of course, with more than a tinge of tinfoil to it, but for a scientist even the most provocative what-if is of little consequence if it can't be calculated. A formula helps focus one's thinking.

The Drake equation, generally accepted, will become the standard for such speculation.

A bit later a new generation of skeptics will propose, in turn, a rival thesis, the rare Earth hypothesis. It holds that intelligent life is actually a scarce commodity, that its development on Earth is due to a rare conflux of factors that aren't likely to be replicable anywhere else. Our world may be the only such data point we'll ever discover. But our existential status will be elevated in this scenario.

Three perspectives, three possibilities, something for everyone, and a curious nonscientist can plunge into the problem without too many qualms. After all, even the professionals are arguing amongst themselves.

As it happens, SETI researchers need help—the raw data they're accumulating have overwhelmed their own computers. There's just too much undigested information to process, and so far they've searched only a small sliver of sky. They want to speed up the work, but the program will be hopelessly hobbled unless they can dramatically increase their data-crunching capabilities.

Someone has an idea—why not harness the power of all those PCs in ordinary homes? Why not enlist all those amateurs who haunt the SETI website, driven by the same curiosity that spurs the scientists themselves? They have no training, but that's okay, their computers can do the heavy lifting.

It's an inspired suggestion. Over a million people download the SETI@Home software and get to work.

The SETI program is passive, designed only to receive signals. That means just watching and waiting while the large dish arrays swivel slowly across the sky. Some researchers aren't patient enough for that cautious pro-tocol; they want to engage in something aggressive. Enough of this slow-motion stuff. We should send our own signal.

Others say it's foolish to expose ourselves; we ought to stay under the radar. Whatever's out there may not be friendly, and we shouldn't be overly eager to find out.

SETI researchers will spend decades trying to find the Wow! signal again, minutely reexamining the region in the Sagittarius constellation they were looking at that August. To their great frustration it will never reappear, but an extraterrestrial origin cannot be ruled out. It will remain the most tantalizing incident in forty years of SETI exploration.

Meanwhile, the expert stargazers who have spent so much time chasing phantoms begin to wonder if the answers they seek will always elude them. As one of them puts it: *the lack of signals is starting to worry many scientists.*

Maybe we're alone, maybe we're not, maybe we'll just never know. But it won't be for want of trying.

14

JUNE 24, 1947: Kenneth Arnold, a private citizen from Boise, Idaho, is piloting his own small plane near Mount Rainier, the highest peak in the area, at 14,410 feet. He's made the trip before (it's the quickest way to get around the Northwest) but this time he sees

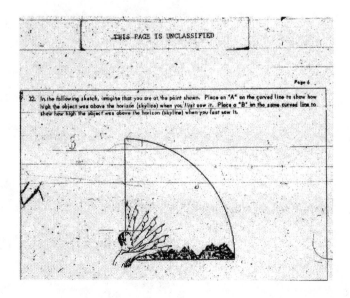

THIS PAGE IS UNCLASSIFIED

Page 6

12. In the following sketch, imagine that you are at the point shown. Place an "A" on the curved line to show how high the object was above the horizon (skyline) when you first saw it. Place a "B" on the same curved line to show how high the object was above the horizon (skyline) when you last saw it.

something funny in the sky, a bunch of things flying in formation. They're not planes—he's never seen their like at any airport in the area. He's concerned enough to report the incident to the authorities.

They want more information. A businessman can't be expected to possess military acumen, or give them hard data, but at the very least they'd like as precise a description as possible.

He mulls it over. They looked like *pie plates skipping over the water*. It's a homey image, inspiring the newspapers to take it a step further. According to them, Arnold saw *a flying saucer*.

That's enough to get the attention of the United States Air Force. It launches a formal investigation—code name Project Bluebook—into what it calls *Unidentified Flying Objects*.

In April 1952 the mainstream magazine *Life* joins the debate. Most of the issue is devoted to the usual popular topics (there's a picture of Marilyn Monroe on the cover) but it isn't a movie star that piques reader interest on this occasion. They go straight to the article titled "Have We Visitors from Outer Space?"

For the first time, ordinary readers are told of

scientific evidence for possible interplanetary visita-
tion; more importantly, they learn that *the Air Force
invites all citizens to report their sightings.*

In 1952 alone more than fifteen hundred people will
do their duty, filling secret Air Force case files with
long, handwritten letters describing sightings from
backyards, front porches, rural areas, cities, convert-
ibles, sedans, and airplanes. The objects are shaped
like discs, cigars, spheres, boxes, and sometimes birds;
they're bright amber, pink, cherry red, white, silver,
and sometimes shiny. The *observers*, as they're called,
fill out follow-up questionnaires developed expressly
for the investigation. They're asked to gauge the reliability of their recollections (*Certain, Fairly Certain,
Not Very Sure, Just a Guess*); they're asked about cloud
formations, weather, wind, and temperature; they're
asked to determine the shape, color, maneuverability,
and airspeed of the objects. They do some drawing.
Finally, they get to the heart of the matter.

*In your opinion, what do you think the object was and
what might have caused it?*

The Air Force diligently acknowledges each submis-
sion (*your interest in reporting the matter and your public spirited desire to be of service are greatly appreciated*). It

concludes, officially, that the writers of the letters represent an impressive swath of American life, from *mystics
to highly educated individuals.* But in the fine print of the
report summaries the word *crackpot* appears with some
regularity. It's probably smart to cover all the bases.

April 20, 1952: Edmund Kogut, age twenty-nine, is at
a drive-in movie in Flint, Michigan, with his wife Shir-
ley. He might be watching *Destination Moon* or *The Day
the Earth Stood Still* or *The Man from Planet X,* all recent
releases. It's still too early for *Invaders from Mars* or *It
Came from Outer Space,* which won't be out until 1953.
But for Kogut the real show isn't on the screen anyway;
he's looking at the *fuzzy balls of fire* that are moving in a
V formation across the sky.

He opens the car door to get a better view. Once his
eyes have adjusted (*the interior car light came on, blinding me momentarily*) he picks up the objects again and
makes some mental notes. They're moving at 1,200+
miles per hour, he estimates, with an ease that seems
unnatural; *the speed and gracefulness which these objects
displayed could not be duplicated by any airplane.* He's
never seen anything like it.

Kogut notes that he's been a licensed pilot for years

and is currently a member of the Naval Reserve. Additional bona fides include an aptitude for subtle discriminations, the sort he first demonstrated in the service (*my aircraft recognition marks in flight training were above average*). The careful drawings he's made indicate a clutch of tiny crosses moving forward, each with wiggly lines radiating outward—a representation of the super-bright heat produced (he surmises) by *skin friction*. He ends his report crisply: Whatever he's seen, it's *not an illusion*.

Working with Kogut's information, the Air Force declares the incident an instance of misidentification: *it is believed that the UFO's were possible bird flights known to frequently fly over that region on their way to the Great Lakes*.

And that's it for the Kogut case.

April 29, 1952: Mrs. Sarah Wilson, eighty-seven years old, sees something unusual while sitting on her front porch in Topeka, Kansas. She describes it as oval-shaped, pale pink, moving back and forth across the sky at great speed. She watches it for about thirty seconds, then it disappears.

Mrs. Wilson isn't going to write a letter, but she's

willing to make a call. Delmar G. Fox, agent, Air Force Office of Special Investigations, takes her information and passes it along to the director of Intelligence, 90th Bombardment Wing, Forbes Air Force Base, Topeka, Kansas.

This one stumps the Air Force. They can't get much more from Wilson, whose skill and technical background are judged *Very Poor*. But it's only prudent to humor the old girl, she's the kind who can make a squawk if she's not happy about something. Further investigation fails to find any conventional aircraft or weather balloons in the area, and no known celestial phenomena can account for the sighting (they've consulted the astronomy department at Washburn University in Topeka just to make sure). They come up with only one hypothesis: *spotlights used by local drive-in theaters reflecting against overcast sky could have produced effect similar to the one observed.*

They owe this one to Mrs. Wilson's descriptive powers.

June 10, 1952: Rose Mary (Mrs. Charles) Hudak, thirty-seven, is sitting in the front parlor of her mother's farmhouse six miles south of Elyria, Ohio. She's looking out a large picture window, dreamily watching a languid country evening unfold.

The stars are bright, even beautiful.

At approximately 8:30 PM, her heart skips a beat; she sees a *large round, silvery white, ball-like object* fly through the air. She's startled enough to call out to the kitchen, but before anyone else can arrive the object has vanished. She's the only one who saw it. She knows the government has been asking around, but she never thought twice about it, not until now. Even a phone call to her sister doesn't calm her down.

Am I crazy I asked myself?

It's a burden to have such information; she almost wishes she'd closed the curtains before it came, that someone else had seen *the thing* (that's as technical as she's going to get). But they want a letter so now she'll have to write one.

Best to begin with a disclaimer. She didn't ask for the honor, after all. *I am a housewife not acquainted with aeronautics or the sciences.* She reports what she saw, how she felt, and remembers to add that someone in Lorain saw something at the same time—she knows because her sister told her so the next day. That's ten miles away. The whole area is buzzing about it.

But best to end on a more subdued note, she doesn't want to seem excitable. *I am not a high-strung or imaginative type of woman and I wanted to make this record of what I have seen.* That's it—a page and a half is all they can ask.

The case is deemed inconclusive, but at least one person in Ohio will never look upwards again without a little bit of worry.

August 1, 1952: Herbert Turner, a thirty-two-year-old salesman, is driving towards East Thirty-ninth Street in Hampton, Virginia. It's about nine o'clock on a warm night; a slight breeze is building, it feels like a thunderstorm is coming. The Florida resident is on a trip with his family (they're visiting his mother, actually, so you might say it's business as well as pleasure). Abruptly, he sees something ahead, something strange, and wonders if he ought to report it. He's not sure; maybe the Air Force will wonder about him. He'd better think about how he presents himself.

He's only on Question One of the report form they've sent him, but already he's second-guessing again. After a moment he crosses out his first choice, *Fairly Certain,* and circles *Certain* instead.

No going back now. Might as well just jump in.

It would have been impossible not to notice while driving as it came into clear view of windshield and also three other occupants saw it at same time.

He doesn't mention that one of the witnesses was his seven-year-old daughter, a minor whose reliability might come into question, but at this point he's just making the initial pitch. It's an old salesman's technique, confidence boosted by bravado; he probably doesn't realize he's setting the presentation up in this way, but old habits die hard. This way he gets his foot in the door, and maybe even makes the sale to himself.

They want to know what he thinks he saw.

He's sure it's not a flying saucer—it looked more like a *huge tracer bullet* arcing across the sky. That's technique too: Surprise the customer with something unexpected, something different from what everybody else is offering.

How many miles per hour did it move?

He hasn't got a clue. Better to admit it, his close will be more effective.

I am incapable of estimating this.

He's almost finished. Just needs to make the pitch one more time. What he saw, he declares, is *like absolutely nothing I have seen before and I have seen many flying planes.*

Someone has put a large question mark next to this statement on the form in the file; it's impossible to determine whether it was the Air Force that did it, or Turner himself.

Another inconclusive resolution.

December 1, 1952: A number of observers near New York City report a single, round, white or amber object floating in the sky. One of them is the pilot of an Eastern Airlines flight to LaGuardia; the others are air traffic controllers and airport personnel who keep an eye on the sky for a living. As professionals, they temper their excitement with information.

All observations placed the unknown in the NW approximately 15 degrees above the horizon on a 300 degree azimuth heading with a slow drift to the south, finally sinking out of sight.

That gives the Air Force investigators exactly what they need. It's got to be the planet Jupiter, appearing precisely where science says it should in early winter 1952.

Undoubtedly, the unknown object is thus explained.

What isn't clear is why so many aviation profession-als don't know the first thing about astronomy.

The deluge of UFO reports has greatly distressed certain constituencies in the government. The whole effort represents a huge drain on resources. It's scien-tifically iffy, and (let's face it) otherwise stable people are suddenly frightened of their own shadows. They ought to be waxing their cars and tending to dinner instead of bolting at each bright light or shifting cloud formation.

The CIA is directed to convene the Robertson Panel in January 1953. Some say its secret agenda is to debunk the very idea of UFOs. Five distinguished physicists are drafted into service. The panel meets, assesses the situation, and issues its judgment: Continued reporting of unexplained phenomena only encourages *a morbid national psychology and harmful distrust of duly consti-tuted authority.*

Translation: People are getting hysterical, they shouldn't be egged on, and it's the government itself that's doing it. Mark something SECRET and you're screaming you've got something to hide. Who knows what that brings out of the woodwork.

Case in point: Mrs. Ingrid Olafson of San Francisco. She first reports seeing a flying saucer while on a bus in 1952. She will spend almost ten years of her life trying to make sense of that fleeting vision.

Her obsession began innocently enough. Soon as she got home she pulled out the binoculars and saw the saucer again, along with four black *birdlike things* clustered around it, with what looked like *little human beings* inside. That made her knees weak, as she put it. She knew it sounded crazy, even her husband laughed at her when he came home, but she made her report to the government in good faith.

I hope you will believe me as I do not tell lies, at least not without having a good reason.

She writes again in 1961, announcing that *since that day I have spent all my spare time delving into the "Flying Saucer" matter.* She's taken an astronomy class in evening school, read all sorts of books, bent the ear of anyone who will listen as she puzzles it out. Science hasn't been much help. She's finally found answers in Scripture, especially Ezekiel, along with insights taken from *the Glorious Koran and the Gospels translated from Aramaic.*

Her twenty-five-page letter outlines an elaborate cosmology that could be characterized as crackpot, though she herself dismisses stereotypes like *beings with long antennae sticking out of their heads.* As she sees it, her conclusions might almost be mainstream. The evidence is there for anyone to sift through, *if you would only use your heads.* God and all his Angels are way ahead of us, *so far advanced in scientific knowledge that we can't comprehend it all.* They're the ones who've been showing up, over thousands of years, just to study us.

As sure as she is of that, she's even more certain of this: They will be back one day.

She signs the letter *Cordially,* though she wonders if anyone's really paying attention.

Another government-sponsored report in 1968 (officially designated the Scientific Study of Unidentified Flying Objects) concludes that no further study of the UFO phenomenon is justified; there's no reasonable likelihood that such scrutiny will advance scientific knowledge. But partisans are prepared this time around. The report is seen as wildly compromised from the start, a deck stacked by known skeptics. It only hardens entrenched resistance on both sides.

Project Bluebook will process over twelve thousand reports of UFOs over the nineteen active years of the inquiry. When it's mothballed in 1970, civilian interest in the unexplained phenomena will continue, though driven underground.

Intermittently, a few scientists on opposing sides of the debate battle on. The profession isn't paying much attention anymore; no one wants a career to founder on some foolish dead end.

In 1968 James McDonald, physicist and proponent of the extraterrestrial hypothesis, makes his case in an appearance before Congress. It's a minor committee, convened by a couple of congressmen who are still interested in the subject. McDonald thinks science hasn't taken UFOs seriously enough. In the course of his testimony he takes a few shots at his prime nemesis, UFO debunker Phillip Klass. It's personal. Klass scorns scientists who take that stuff seriously, and he's mentioned McDonald by name. He's also offered $10,000 to anyone who can provide proof of extraterrestrials. So far, no one's claimed the money.

Klass says it's simple: Unidentified objects seen from aircraft (for example) are natural phenomena, *plasmoids*, or gassy clouds trailing behind the plane, drawn by its charge. Scientists call it the *Coulomb attraction model.* What pilots see, the bright fuzzy balls, may seem otherworldly, but aren't.

McDonald fires back: *the interesting UFO reports do not involve hazy, glowing, amorphous masses.* He's said for years that serious investigation ought to address the machinelike mechanisms that have never been explained, all those discs with ports and domes that observers have described in compelling detail. It's bad science to consider only the obviously kooky stuff and ignore the more puzzling phenomena.

That said, the plasmoid argument can be refuted by doing some basic calculations: *Since the surface charge density omega will satisfy $E = 4\pi \times \Omega$, each object will then hold a charge $Q = R^2E$ (esu) where R is the object radius and E is taken as 20,000 V/cm = 65 esu/cm.*

From there, anyone who knows the physics can see that Klass's plasmoid theory is utterly improbable.

For the rest of us, McDonald says it more simply: *he is using arguments that would collapse if he were to try to put numbers into them.*

But it's too little, too late.

At this point the average person, pressed to adjudi-
cate the issue, is as much in the dark as Mrs. Olafson.

15

◎ Thought Experiment

Is it possible for an ordinary person to acquire the mindset of a scientist?

It's an ancient challenge; even the great Archimedes was probably misunderstood by members of his own family. But let's assume that everyone has the potential, even someone as possibly resistant as yourself. (It isn't relevant, in this instance, that you've never taken statistics, or that you let your mother make the volcano in that fifth-grade assignment.) Relax, there won't be a test at the end—in an experiment, you control the parameters.

But first, you'll have to make some adjustments in the way you think about language. For instance, scientists don't use the word theory *as the rest of us do—it's not a wild guess or a favorite belief, but a general principle that explains natural phenomena and is founded upon empirical fact. And it has predictive power.*

If you understand that, you're halfway there.

Now select an object or substance near at hand and examine it carefully. Is it solid, liquid, or gas? Is it rough or smooth? Does it break rather than bend? How fast does it fall?

Repeat the process with other objects. You may, if you wish, take notes; putting important observations on paper will free up more brain matter.

Now take another look around the house. Do you keep stirring rods, suction bulbs, beakers, and corks close at hand? Do you have any equipment at all? (Remember to look in the kitchen.)

If so, you can proceed to the next phase—it's time to identify a phenomenon that you'd like to investigate. If you can't think of anything, try these sample challenges: Do metals rust at different rates? How does fog form?

Once you've got the idea, it's time to focus on procedure.

Do you have a hypothesis? Have you designed an experiment to test your idea? Will your data allow you to draw conclusions?

It's possible you don't yet feel ready for advanced work, and that's okay. Proceed at your own pace. In the meantime, ask yourself this: Are you interested in the world around you? Can you express that curiosity in the form of questions?

If so, congratulations: You're starting to think like a scientist.

The Royal Institution, London, December 1860: Michael Faraday, premier experimentalist of the nineteenth century, gives an electrifying lecture to young people on The Chemical History of a Candle. At his urging English scientists have delivered Christmas lectures since 1827; the annual series has thrilled generations of children who hadn't any idea how exciting *experimenting* could be. Each series explores, in depth, a single scientific topic, using demonstration, skillful explication, and dazzle to make the fundamental principles of science come alive.

It's easy enough to draw children to talks about electricity or soap bubbles; Faraday's genius lies in his realization that The Chemistry of Coal, The Voltaic Battery or Combustion can, in the proper hands, prove every bit as enticing.

A bright boy born into hard straits, he got most of his early education from books he read by happenstance while apprenticed to a bookbinder (he especially liked the consuming *Encyclopædia Britannica*, then in its third edition). He couldn't go to school, but he fervently wished to be a scientist. His gifts were recognized, for-tuitously, and now as an established master he works to spur the inchoate ambitions of others.

Faraday's lecture on the candle becomes an immedi-ate classic. He's found a way to dramatize a chemical process—as the series unfolds, the candle itself becomes a character, its behavior as involving as anything the boys and girls will find in Dickens. But Faraday him-self is also a phenomenon, the bubbly embodiment of the delight he takes in the ways of the natural world. *I claim the privilege to speak to juveniles as a juvenile myself,* he tells them early on, all the while insisting that they work as hard as he does. *I want you to put these different facts together in your own minds; this is a good experiment for you to make at home.* Genuine education can't be con-fined within the arbitrary limits of a lecture.

In the end, he leaves his young charges with one abiding lesson.

I hope you will always remember that whenever a result happens, especially if it be new, you should say "What is the cause? Why does it occur?" and you will, in the course of time, find out the reason.

◦ ◦ ◦

American children of the 1950s have their own science enthusiast in Mr. Wizard, a phenomenon of the
emerging medium of television. For fifteen years kids
make an appointment to watch Mr. W, a low-key show-
man devoted to *the magic and mystery of science.* The half-
hour episodes are shot in grainy black-and-white with
few apparent production values. There's nothing to see
but the science itself, and in the '50s that's enough.

Each show features a young assistant who works
with Mr. W and parries his constant question: *Why?*

Rita, in ponytail, pleated skirt, and knee socks,
doesn't like being wrong; she's happiest when she
comes up with the answer right away, but keeps on

trying if Mr. Wizard says *no, no,* and once again asks *why*?

Occasionally he tries to trick her, as when he announces that a substance called *phlogiston* is responsible for the little fire she's just lit. But she's having none of it, she's never heard of that whatever-you-call-it, and she's been paying attention. She's got him. He admits that phlogiston is something no one's taken seriously since the 1700s, and Rita learns that authority isn't always right.

Alan, in a child-sized oxford shirt and pressed slacks, is happy simply to watch Mr. Wizard's wonders. Sometimes they make sparks and sounds out of the gizmos he's built, and sometimes they measure things or study insect locomotion. It's all the same to Alan. He laughs easily, likes clues, and gets the pronunciation of *piezo* right on the second try.

Whatever they're investigating, Mr. Wizard never lets the kids off with a shrug; he wants them to understand. *It's all about molecules*, he tells them, in just about every show.

In one classic episode Rita and Mr. Wizard explore how *adhesives* work. At his direction she makes a gluey paste out of substances found in the kitchen; in short order they've stuck two pieces of wood together and measured the strength of the bond. She learns that the runny stuff she's cooked up (*a liquid*) will seep into the tiny spaces between the two boards, and as it dries into *a solid* the pieces will prove almost impossible to tear apart.

It's molecules again.

At the end of the show Mr. Wizard makes stilts for her, using her glue to fasten the wooden blocks and sticks together. He encourages her to test them. She's nervous, afraid she'll fall (that might be worse than being wrong), but she gives it a go and wobbles around for the camera as the closing credits roll. She's smiling.

Mr. Wizard has taught her something else: Trust in science, it will keep you upright.

The program is believed responsible for a surprising spike in applications to graduate science programs in the 1960s and '70s, the applicants all those kids at home—glued to the screen—who want more of the magic they once found in one word: *Why?*

16

September 2003: Clarence Birdseye—father of frozen foods, inventor of the freezer—is inducted, at long last, into the Food Engineering Hall of Fame. It's an honor that's overdue.

Birdseye is posthumously cited for work he did in the 1920s, developing *innovative food preservation processes* that put vegetables, fish, and meat on the table at any time of year. To the homemaker, it was magic;

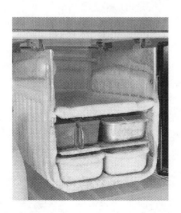

to Birdseye, it was chemistry, tempered with trial and error.

By using calcium chloride and ammonia in his freezing device, he found that foods could be *quick frozen* to 0°F. Such speed radically constrained the growth of ice crystals in delicate cellular structures, preventing tissue damage, which meant that everything from fruits to carcass meats could be put away for the future.

The science itself seemed complicated to most people, and Birdseye didn't advertise its details. Almost no one realized that calcium chloride is just salt. He didn't emphasize the personal details either—the homemaker didn't need to know he'd been inspired by time spent among Eskimos, or that he personally preferred frozen caribou, polar bear, and beaver tail to more domesticated dishes. Most of his customers hadn't taken college courses or traveled, as he had, but everyone liked to eat. He'd make a fortune if he kept it simple.

Bottom line: Birdseye's food tasted fresh after months in the fridge.

To some inspired by such discoveries a revolution in the national diet is the least of what the new powers portend. At root, the idea of such preservation has opened the door to something far more profound—a way to reverse nature's cycle of decay. Science has been searching for such power for a very long time. It's shortsighted to apply the process only to green beans or peaches out of season; other types of tissues might respond positively to the new techniques.

Who knows what else can be frozen?

January 1967: James Bedford, nominally deceased, is carefully lowered into a bath of liquid nitrogen at -196°C. He is thereafter known as *the first man ever frozen*. (History will note that the first attempt to freeze a woman, two years earlier, didn't work.)

Bedford has put his faith in cryonics, a new field that most conventional scientists shun. It's one thing to study the effect of very low temperatures on living tissues, as cryobiologists do, but altogether something else to extend that interest to the macabre. There's also a dispute as to what actually constitutes death, and whether the body, if frozen, can retain its viability until resuscitation. Bedford, unfortunately, doesn't have the luxury of waiting for science to work it all out.

Terminally ill, he has to take the plunge as soon as his vital signs cease.

Bedford's wife has opted for cremation upon her own death, so the first *cryonaut,* as he's dubbed, will be *making the journey into tomorrow alone,* according to company literature.

But cryonicians in the 1960s are confident that someday a widespread "freezer program" will be in place.

Clients of the start-up have a choice, head or whole-body preservation. It's cheaper to cryoprotect the brain itself, and probably smarter—why take a damaged carcass on the trip? Why not just preserve the 5 percent of you that really counts?

The goal of *neuropreservation* is ice-free preservation of the human brain. Cryoprotectant fluids are channeled into and between the body's cells to stop icing—*the solution becomes a glassy solid, locking all molecules in place.* (It's molecules again.) In the future, when science is ready, the patient/client will be restored to life by growing a new body around the brain. That will depend on the development of tissue regeneration technology, as well as the capability of repair at the molecular level—implicit in the developing science of *nanotechnology*—and advances in resuscitation research.

It's a tall order, as the company itself admits: *at the present time the technology required for the realization of our goal far exceeds current technical capabilities.*

But the prospect of coming back to life, even centuries hence, is powerful enough to induce hundreds of people to sign up for the service.

Significant unknowns remain, among them the unsettling possibility that the neural preservation process may cause damage of its own, an offshoot of twentieth-century limitations.

One's personality may be altered, memories may erode, the intellectual capabilities of the mind may be impaired; conceivably, one's brain might even be reduced to an insensate lump, the neural equivalent of freezer burn.

Or the whole body, should it be preserved, may end up a sodden mass.

There's just no way of knowing.

As it happens, twenty-five years into his suspended state, Bedford is scheduled to be moved into a new, improved storage canister. Science hasn't yet found a means to revitalize him, but it can offer incremental advances at the margins.

He must be carefully unwrapped before transfer. Cryonic technicians—many of them slotted for eventual cryopreservation themselves—have a vested stake in his well-being, and some professional pride. They're nervous, even worried. Who knows what they'll find inside the plastic sheeting?

Bedford emerges. Hallelujah!

He looks great, no ice crusting anywhere, just a body waiting patiently for someone to awaken it.

In 2004 sixty-one scientists, intrigued by the effort, write an open letter for the company *to whom it may concern. Cryonics is a legitimate science-based endeavor.* They aren't afraid to look foolish; they know the history of science is replete with achievements that once seemed utterly fantastical.

But the average customer may respond more readily to the pitch first employed by the company in 1964: *Why not try for a life in the future via freezing?*

17

○ *Thought Experiment*

Imagine you're part of a culture that has lost all sense of scale. There's only Extra Extra Large, Jumbo, and Humongous; bigger and faster is always better (only artists or occasional architects think that less is sometimes more). If just a few million people watch a TV show, or ten thousand buy a book, it hardly moves the needle in this disoriented day and age.

Travels that once took weeks or months now take only hours; messages arrive instantly, more agile than thought itself. The phrase "larger than life" has become normative, while expressions like "I'm not myself today" describe an experience of incommensurability that's now increasingly common.

It's a challenge just to get one's bearings.

To many thoughtful people, of all ages, the world seems strangely out of sync. Some find the disjunctions interesting, others are disconcerted, many are simply overwhelmed.

It's all relative, unless it isn't.

Question: Is a centuries-old mania for measurement finally exhausting itself?

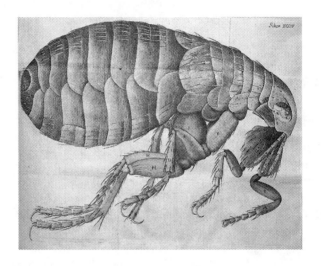

London, 1663: Robert Hooke, young curator of experiments for the Royal Society, assistant to Robert Boyle, has just begun to play with a novelty known as *the microscope*. He's fascinated by its powers of magnification. By peering into two glass lenses attached to a cardboard tube, one can see a thousand times more minutely than with the unaided eye. The senses themselves are weak; when assisted by instruments, *there is a new visible world discovered to the understanding.*

It's as if a fog has lifted, and the Earth is suddenly alive with creatures hardly suspected before.

Hooke seizes on the world of the everyday; *the most vulgar instances are not to be neglected.* Common gnats and fleas reveal an unexpected magnificence, a beauty of form that takes him aback.

He calculates that a million mites can be contained in a cubic inch.

A slice of cork reveals astounding numbers of tiny units he calls, for the first time, *cells.*

The designs of nature are mind-boggling.

He subjects his own hair to scrutiny. He can't see the DNA molecules, that won't be possible until more powerful microscopes, capable of millions of magnifications, are devised in the twentieth century. But he does confirm that the strands are *prismatical, neer round,* and very much worth the pulling.

It's the least one can do for science.

He observes something very small crawling across a book he's just opened; in a moment, under the microscope, it reveals eight crablike legs with claws flailing in unison.

It seems *my little Objects* are everywhere.

◦ ◦ ◦

King Charles II asks for a personal demonstration of the new *microscopy*; like everyone else, he's astonished by what he sees. He encourages the young naturalist to gather his intricate drawings, with their dazzling descriptions, into a book.

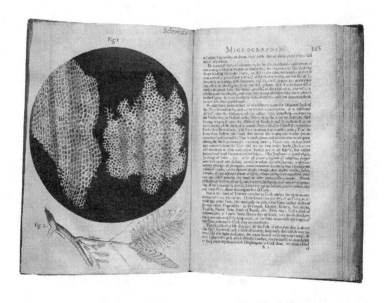

Micrographia becomes a best seller, a catalogue of wonders that keeps even the redoubtable Samuel Pepys up till all hours. Who would have thought the world underfoot could possibly be so engrossing?

Hooke himself grapples with the vastness of Nature as it appears to the seventeenth century, with its telescopes, microscopes, and sobering capabilities.

Some parts of it are too large to be comprehended, and some too little to be perceived.

And humanity itself is caught betwixt and between.

December 29, 1959: Physicist Richard Feynman delivers an informal talk at Cal Tech that will prove to be one of his most prescient. He's interested in *the problem of manipulating and controlling things on a staggeringly small scale.*

It's actually a thought experiment—science in 1959 can only conjecture about such possibilities. But we know the laws of physics, and what should work in principle.

To frame the subject in the proper spirit, Feynman poses a question: Why can't all twenty-four volumes of the *Encyclopædia Britannica* be written on the head of a pin?

In principle, no problem.

Assume the pin is one-sixteenth of an inch in diameter; if the text is reduced twenty-five thousand times, so that you're talking about letters as small as atoms, you should easily be able to shrink the entire *Britannica* to fit. The resulting text could be read only by using an electron microscope.

If that seems fantastic, consider that even the tiniest cell in the human body stores vast quantities of DNA information, ready to be decoded as necessary.

Feynman admits that such feats aren't yet practical, given the technology of the day. Electron microscopes are relatively primitive. Why not make them a hundred times better so that we can isolate and work with individual atoms? Computers (which now fill whole rooms) should be shrunken until they're submicroscopic in size. And that's just the beginning.

Ultimately, we can manufacture *infinitesimal machines* that maneuver on the atomic level. Medicine would be transformed—a person could *swallow the surgeon* and set it to work on heart repair. And who knows what kinds of new materials and products might be built using such powers of precision.

To those who don't know physics, and to some who do, the whole idea seems pie in the sky. But there's no payoff in science if you play it safe.

Feynman doesn't yet have a term for it, but he's talking about the future, a still distant vision in 1959. In decades to come scientists will routinely work on the level of a nanometer, a unit one-billionth of a meter in size. That's about the width of five carbon atoms.

For Feynman, trapped in the 1950s, such efforts can't come too soon.

We are not doing it now simply because we haven't yet gotten around to it.

September 17, 2002: Congress convenes the first Senate hearing on *nanotechnology*. Officially, it will be billed as a meeting of the Subcommittee on Science, Technology, and Space, under the Committee on Commerce, Science & Transportation. The catchall categorization is fitting, for nanotechnology is an area that wantonly blurs boundaries, blending elements of physics, chemistry, biology, materials science, medicine, information technology, national security, agriculture, and manufacturing. And that's just the field in its infancy.

The hearing is being held to announce the 21st Century Nanotechnology Research and Development Act. The committee members have already studied up on the subject; they've got a whole nation to educate, starting with some fellow senators.

Ron Wyden of Oregon opens with a disarming admission: *in coffee shops and senior centers this afternoon Americans are not exactly buzzing about this science.*

But that's about to change. Nanotechnology, he explains, will utterly transform the way they, their children, and grandchildren live, with greater repercussions than the Industrial and Computer Revolutions combined.

George Allen of Virginia adds the folksy touch. Nanotech, he explains, involves manipulating atoms and molecules a lot smaller than the width of a human hair; *of course, you would have to look at that under a microscope—you probably could not see it with the naked eye.* There's no indication that he's done so himself, but he knows what really little is, and bets that it's bound to be good for Virginia.

Wyden catalogues the benefits from nanotechnology foreseen in the next twenty years, given the necessary research stimulus. First, the small stuff: Pants will be unbelievably stain-resistant, windows will be self-washing as nanoparticles scrub the surfaces on their own. More remarkably, a device the size of a sugar cube will hold all the information in the Library of Congress. In medicine, tiny mobile bulldozers will unclog blocked arteries, and anticancer drugs, targeted on a molecular scale, will be delivered to damaged cells directly, killing only the cancer and leaving the healthy tissue alone.

Nanoscience will extend the human life span, and possibly even spur damaged tissue to regenerate.

Military applications abound as well. Nanoparticles will make a soldier's body armor virtually invisible, and such particles could also begin to heal wounds on the battlefield. Nanobadges can serve as biosensors, so finely tuned that even a floating atom or two of any noxious substance released by a terrorist will trigger an alert.

Spacecrafts could be the size of mere molecules.

And it's all based on solid science.

Witnesses add additional perspectives. The businessmen cut to the chase: Nanotech will be *a trillion dollar global market* in a decade. The number of nanotech start-ups has multiplied tenfold in just three years. The U.S. Patent Office can't handle the multifaceted nature of nanotechnology; it's too twentieth century, and it's getting in the way.

Alarmingly, the European Union, Asia, and India

are all making massive investments in nanotech that threaten to leave the United States in the dust.

And, finally, Americans don't know enough about science, and its study will have to be incentivized. Science for science's sake is great, but only Pollyanna believes that's enough of an inducement.

You have to tell them there's major money to be made.

Suddenly, everyone is salivating, expert and non-expert alike.

18

WHAT IF YOU'RE tired of making glue, or playing with robots and rockets, and would rather think about living things?

You might take as a model the friar Gregor Mendel, who found great inspiration in his garden.

In 1856 he began to ruminate upon peas—*Pisum sativum*—and over the next eight years grew and studied twenty-eight thousand pea plants. He wanted to understand the mechanisms of heredity. He was mathematically inclined; the vague notions of "blended" traits previously thought to account for the phenomenon were intolerably fuzzy to him. Anyone could see variations in successive generations of plants (with peas, some stems were long, others short; some pods were green, others yellow), but no one had thought to determine their statistical relations. Without rigorous, controlled experimentation you'd never get further than "more or less" likely, which means you might just as well guess.

Granted, that had always seemed good enough.

Mendel paid dutiful homage to the work of Kolreuter, Gartner, Herbert, Lecoq, Wichura, and all the others who'd devoted their lives to the dogged study of plant life.

But Nature had to be more predictable than that.

Eventually, Mendel's work would establish the basic laws of genetics, though he wouldn't be recognized for that achievement until long after his death. He must have been disappointed when his lecture "Experiments in Plant Hybridization" provoked no discernible interest at the Natural History Society of Brünn in 1865. But the word *gene* hadn't even been coined yet; he could hardly expect his contemporaries to understand the concept. Besides, he'd already failed his teacher certification test, and that had to have undercut his credibility.

Like many scientific pioneers, he knew he'd achieved something *(my time will come)* but he wasn't quite sure what.

Such reckoning is what the future is for.

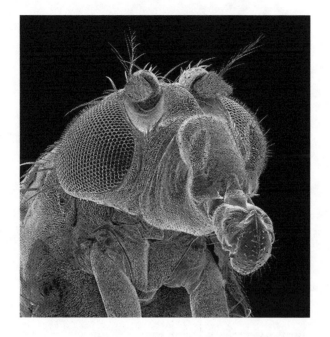

You may not have the time or patience to grow peas, especially if the significant work has already been accomplished. But there are other research opportunities at hand.

It might come as a surprise (if you haven't got your high school biology book) that the pest you swat away in summer, the lowly fruit fly, is actually *Drosophila melanogaster*, a scientific superstar.

It's been studied exhaustively for almost a hundred years. And still it keeps secrets.

Drosophila is one of a small number of "model organisms," the elite of scientific research. It has a life span of just two weeks, is readily found around food, mates with abandon, and produces hundreds of offspring in its abbreviated time on Earth. It can be easily studied in the laboratory or, with a little preparation, in the home.

Curiously, Drosophila also exhibits a propensity to fly towards the light. This may suggest a spiritual capacity, but science isn't willing to venture into that domain. Such musing must remain, for experimental purposes, irrelevant.

Instead, fly devotees stick to the facts, collected in a repository available to anyone. (Freely sharing one's findings is a basic tenet of scientific research.) The University of Michigan offers a handy tutorial on the subject:

> *The common fruit fly is normally a yellow brown (tan) color, and is only about 3mm in length and 2mm in width (Manning 1999, Patterson et al. 1943). It has a rounded head with large, red, compound eyes; three smaller simple eyes; and a short antenna. Its mouth has developed for sopping up*

liquids (Patterson and Stone 1952). There are
black stripes on the dorsal surface of its abdomen,
which can be used to determine the sex of an
individual. Like other flies, Drosophila melanogaster *has a single pair of wings. Larvae are*
minute white maggots lacking legs and a defined
head (Patterson and Stone 1952, Patterson et al.
1943, Raven and Johnson 1999).

Of course, that's just what one sees on the surface.
Drosophila also possesses four chromosomes (first
crudely mapped in 1913) and its simple genome offers
scientists a manageable way of exploring genetic complexities. Over time, as the sequence and functions of
the fly's twelve thousand genes are established, science
will edge closer to understanding the minute workings
of still more challenging species.

Such as human beings.

Hundreds of scientists are even now running
sophisticated experiments with Drosophila as the test
subject. There's something for everyone—aggressive
behavior in Drosophila, embryonic muscle patterning,
cell adhesion, circadian rhythms, molecular genetics
of meiotic recombination, retinal determination and

degeneration, organogenesis and tissue architecture, asymmetric RNA localization . . .

You might as well add your own experiment to the list.

You can purchase the necessary equipment almost anywhere; pay particular attention to ads touting *long-lasting economical propagation media* to save on start-up costs.

It's not just a science, it's an industry.

If you're nervous about plunging in, without formal preparation, you might want to enroll in the Drosophila Genetics and Genomics course soon to be offered in Chengdu, China. One of its guest instructors has written a canonical text known informally as the Drosophila bible (an apt description of a foundational work more than fourteen hundred pages in length). It's an opportunity to learn from an international authority, and get some travel in at the same time.

For those whose interest in Drosophila is strictly begrudging, science offers tips on such vexing issues as infestation.

The most important thing to do is get rid of the food they're growing in. I'd keep fruit out of the house altogether

until they were gone, unless you can refrigerate it. Then it's
just a matter of waiting for the adults to die.

But that's common sense, a nonscientist could come
up with that. Surely experts know something the rest
of us don't.

I just put a dab of detergent in wine (they love it), and
pour it into wide-mouth containers such as beakers or coffee
cups. Flies will place their weight on their front legs on the
surface of the liquid in order to drink from it. The detergent
breaks the surface tension, so the flies fall in and drown.

That's more like it. We already knew that Drosophila
has a mouth designed expressly for drinking. Now
we've learned it's something of an oenophile as well.
All that remains, perhaps, is to meditate upon how and
why strengths tip into weaknesses and sometimes self-
destruction.

But no one's about to call Drosophila a drunk. That
kind of judgment exceeds the warrant of science.

March 26, 2008: Richard Lenski of the Department
of Microbiology and Molecular Genetics at Michigan
State University submits the results of a twenty-year
laboratory study to the National Academy of Sciences
for peer review. If the paper passes muster it will be

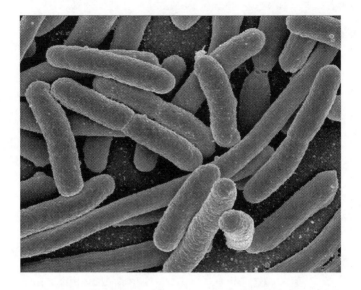

published, and the scientific world will dive at last into the weeds of Lenski's remarkable long-term evolution experiment.

Three months later, when "Historical contingency and the evolution of a key innovation in an experimental population of *Escherichia coli*" finally appears, nonscientists will be able to read and debate the findings online as well.

Of course, many of them won't want to get anywhere near E coli.

◌ ◌ ◌

The experiment is an example of science at its most fundamental. Lenski and his colleagues have taken a hugely complicated problem and radically simplified it, using one of nature's most basic life-forms as their experimental focus. There are five nonillion (5×10^{30}) bacteria on Earth at any given moment, more or less, and E coli is a particularly suitable subject of study. Most of its variants are benign, it's easy to grow, and its genetic makeup is well understood. Lenski and his colleagues plan to track the E coli over time and watch what develops. The bacteria themselves will do most of the work.

Beginning in 1988, he takes two ur-bacteria (known more formally as *ancestral clones*) and develops six initial samples from each. The twelve identical populations are cultured in a mixture of glucose (an energy source E coli gobbles up) and citrate (a source it lacks the mechanisms to metabolize). As the bacteria propagate, one generation every six days, billions of mutations are expected to occur in each population. At some point, Lenski and his group conjecture, it's possible that an E coli strain may develop a more diverse palate.

It's essentially a recipe for accelerated evolution, on a scale measured in years rather than eons.

Who knows what will happen?

Every five hundred generations, or seventy-five days, population samples are frozen at −80°C in the ultra-low freezer, with 1 ml of glycerol added as a cryoprotectant. The frozen strains can be thawed and reanimated at any time; the bacteria are simple enough and hardy enough to survive the cold storage without undue damage.

Such samples will give the experimenters a "frozen fossil record."

Throughout the experiment myriad protocols must be followed precisely.

Transfers should be made 22–26 hours after the previous day's transfer; Label 12 flasks as A+1 through A+6 and A-1 through A-6. (Check flasks and beaker tops for cracks.) Add 9.9 ml of DM25 to each flask. Propagate 12 cultures by transferring 0.1 ml from previous day's cultures into fresh DM25. Incubate the new flasks in the shaking incubator at 37°C and 120 rpm . . .

In addition to putting number stickers on top, use a blue marker to write numbers and strain identifiers on the vials; record the strain numbers and identifications in the lab notebook . . .

For two decades nothing unusual occurs; the E coli thrive, in more or less the same way, over tens of thousands of generations. Lenski's lab assistants prepare and catalogue over twenty thousand different cultures.

But in generation 31,500, *a weak Cit+ variant* emerges in one of the twelve populations.

Bingo.

A strain of E coli appears to have taken to a new menu, if ever so tentatively.

Within two thousand additional generations the variant develops robust citrate-sucking capabilities, and a massive population expansion ensues. The new E coli have apparently *evolved to exploit more efficiently the abundant citrate in their environment*. And thus rise to numerical dominance.

But a few of the E coli, perhaps stubbornly, remain partial to the glucose, leading to two variants of the bacteria residing in the same agar plate.

That's even more worthy of note—increased diversity is a predictable consequence of the evolutionary process.

Using the frozen specimens, Lenski tests his hypothesis by repeating the process dozens of times from various start points.

Once he's replicated the experiment, he's ready to announce his findings and join in the widespread wonderment at *this fascinating case of evolution in action.*

A fractious online debate at a science site begins immediately upon the announcement. Antievolutionists decry the results and express heated skepticism.

"If the bacteria changed, it was because God willed it." "Sorry, there is no substitution for the Holy Bible!"

These *cretards*, as they're dubbed, are faulted in turn for their ignorance of basic science.

"You would have trouble adding vinegar to baking soda unsupervised." "It's clear that you're involved in worldview defense." "I recommend an introductory university course in genetics, or for that matter a high school biology book." "I'm tired of the creationists claiming this never happens."

One can find disinterested exchanges amidst the fireworks, most of them marveling at the achievement itself.

"I forgot how much unexpected fun like this can come from coli." "This is the most excited I've been about biology in a while." "What I really want to know is what is the necessary sequence of mutation which preceded Cit+ along that Ara-3 line?" "If they provide a gene sequencing paper that

is verified by others, then yes, he will probably get the Nobel Prize. Certainly not until then though."

The larger frenzy only abates when one of Lenski's collaborators joins the discussion. He apologizes for coming late to the party, and encourages everyone to read the original paper.

If you are still puzzled, please post here, and I will see what I can do to help you out.

Civility, meted out in small doses, seems to have a calming effect upon this particular population.

19

○ *Thought Experiment*

Imagine yourself in the lab. Would your work be 9–5, or 24/7? Could you share it with others, would they even be interested? How long could you keep going, if you suspected that only some stranger in the future would appreciate your efforts?

It must be easy to get lost in the minutiae of the day-to-day, to find oneself inured to the wonder, to believe you're just chipping away at small parts of large problems as you label a flask for the twenty-thousandth time. You must lose hope of ever experiencing that eureka moment, or fear that all the big questions have been settled (or abandoned) already. You worry that you're pursuing a dead end. You despair at debating contentious issues over and over again—the atom bomb, when life begins, stem cells, climate change—with people who don't speak the same language. You wish you could travel to a simpler time, and argue about aether, or gravity, or whether the steam engine (as Thoreau feared) would spoil Walden Pond.

But of course nothing as audacious as science has ever been simple.

Now imagine an alternate reality. Imagine that you approach each day with undiminished interest, and conduct your work with diligence and care, and trust that your efforts (sure to be superseded) will be worth every sacrifice if just one new piece of the puzzle is slipped into place.

You believe that working in such circumstances isn't a burden, it's a privilege.

Now imagine that ordinary people, the ones who have other kinds of jobs and interests, watch you from afar and wonder—what would it be like to spend one's time thinking about something so much larger than oneself?

Question: Would you have anything to say to them?

May 31, 2007: James D. Watson, living emblem of the double helix, is given two DVDs containing his own distinct genome, or personal genetic blueprint. It's just the second such individual genome ever deciphered. Watson has been selected for the honor as co-discoverer of the molecular structure of DNA, the 1953 achievement that revolutionized humanity's understanding of life itself.

I am thrilled to see my genome, he allows. It's taken many people two months and a million dollars to do the decoding, but Watson isn't shy about his worthiness. After all, no one would dispute his own matter-of-fact assessment of the work on DNA.

Francis Crick and I made the discovery of the century.

In the 1950s, according to Watson, the primary challenge of biology was to understand gene replication, and just how genetic information was transmitted from generation to generation. Miescher, Mendel, Avery, Franklin, Wilkins, Chargaff, and Schrodinger had done crucial preliminary work, in the incremental rhythms of scientific discovery, but science had only general notions about how the specific mechanisms might operate. Watson and Crick had an intuition—*it was obvious that these problems could be logically attacked only when the structure of the gene became known. This meant solving the structure of DNA.*

Their ensuing discovery, fueled by the construction

of various cardboard and wire models of molecules, would become one of the great epiphanies of science.

On a personal level, Watson found the double helix formation *unbelievably interesting*, meaning it presented ample possibilities for good science. But there were larger implications as well.

We knew that a new world had been opened and an old world which seemed rather mystical was gone.

The secret of life could be deciphered, down to the nuts and bolts.

Watson would become the first director of the Human Genome Project—the Herculean effort to unravel the entire genetic code of the human species. (Francis Collins would see the project to completion in 2003.) The reference prototype was constructed from a composite of anonymous donors; it took ten years and billions of dollars to specify the three billion DNA base pairs and the twenty-four thousand individual genes of the human species.

Now that the sequencing has been established, science must begin the painstaking work of determining the functions of each gene.

Some predict that the decoding of this *instruction*

book will prove the major scientific discovery of the twenty-first century, and possibly all time.

Science now knows that every human has 99.9 percent identical DNA, that humans have just twice as many genes as a fruit fly, that 50 percent of the genes of humans can also be found in a banana, and 80 percent in a mouse. It's increasingly clear that all life is part of an interwoven web, with the gene itself at its center. Old distinctions between species and organisms, cultures and races are certain to fade, Collins predicts, as the gene becomes the new focus of interest.

And humanity will undergo another radical shift in its own sense of itself.

Within fifty years, scientists say, diseases may be cured at the molecular level before they have a chance to develop. And in the near future, anyone with $10,000 will be able to order his or her own genome from private industry. Labs all over the world are rushing to grab a stake in that gold mine.

Those who don't have the money will remain hostage to surprise, old-fashioned as that will surely seem. The ethics of the genomic enterprise have hardly begun to be explored.

Watson has offered his own genome to researchers, all but the part concerning the *apolipoprotein E* gene, which is linked to Alzheimer's. He'd rather not know about that.

Soon enough, other people will have to make the same kind of decision.

Watson was just twenty-five when he did the great work on DNA; now, he's almost as famous for disputatious comments made later in life. He believes in the possibility of a genetic underclass, a group of unfortunates whom heredity has dealt a bad hand. Some people are stupid, which could be thought a disease (dumb genes); some (especially dark-skinned peoples) have unusually active libidos; there may be something to the assertion that women don't have the genetic capacity to do top-flight scientific work. He'd like to see the genomes of successful people decoded; he's also interested in the genetic makeup of psychopaths. He'd like to make all women beautiful, through genetic manipulation. Most of all, he'd like to see disease understood and eradicated.

The basis for life becoming better is not spiritual values or anything—it's knowledge.

Watson's genome may present opportunities for research into the genetic basis of inflammatory beliefs.

But culture and environment also play a role in shaping individuals, as geneticists readily concede. To learn even more about Watson from a nongenomic perspective, one need only consult the James D. Watson Collection in the Cold Spring Harbor Laboratory archives. (As former director of the institute, Watson is widely credited with creating a world-class research facility.) The archives constitute a comprehensive repository of every extant piece of information, possibly every piece of paper, related to Watson in any way.

Most prominent is the scientific and personal correspondence—handwritten and typed letters, carbon copies, postcards, and notes regarding social activities and travel plans. Also letters from his father dispensing advice.

One can sift through manuscripts and typescripts, teaching files, personal gifts, memorabilia, scientific reprints, Watson's curriculum vitae, and passports.

Or a collection of his Christmas cards.

Smaller items in the archive include autograph requests, declined invitations, contracts, book reviews, articles, newspaper clippings, congratulatory letters, and telegrams.

And one can pore through his handwritten graduate school notebooks.

For the moment, he's the most extensively documented human being on Earth, each bit of DNA and every Post-it available for researchers to examine. But his case may only be a harbinger of what could become a cultural expectation of utter transparency for everybody.

That's for the future to determine.

20

ELSEWHERE A COMPETING vision of the future is lying in state for all to see.

Tourists with radiation detectors can now visit Chernobyl, the site (some say) of the ruins of science. Many of its visitors hadn't yet been born on April 26, 1986,

when nuclear reactor 4 exploded and sent plutonium, iodine, strontium, and such streaming over much of the northern hemisphere.

The effect of the blast exceeded that of Hiroshima by a factor of 400.

Caesium-137 (a nuclide with a half-life of thirty years) settled into the soil and forests surrounding the site. Rivers and reservoirs were irradiated as well. Contaminants floated into the food and milk of children.

Eventually, an *exclusion zone* was created, a no-man's-land of nineteen miles buffering the most intensely radioactive area. The reactor itself was buried within a concrete structure known as The Sarcophagus.

It would prove harder to contain the fear and apprehension unleashed by the event.

Ordinary people the world over panicked. They lacked expertise in radiology and couldn't assess the risks for themselves. They didn't trust the authorities. Science suddenly seemed malevolent.

But only science could make them even halfway whole again.

Epidemiological studies will be conducted in the affected areas for decades. Scientists are monitoring increases in breast, lung, stomach, uterine, and prostate cancers among first responders and cleanup personnel, as well as a dramatic (and statistically aberrant) uptick in radiation-induced thyroid cancers in the children of Chernobyl.

Medical personnel have achieved a heightened understanding of radiation medicine, and they're gaining insights into the treatment of sociopsychological stress.

The Swiss government has set up a website cataloguing the long-term consequences of Chernobyl. Recently, it added a section called Experts' Opinions at the request of concerned readers.

Scientists are tracking genetic defects in children exposed in utero, as well as chromosomal aberrations in resurgent plant life. And experts are vigilant about monitoring *radionuclide migration*.

It seems even the radioactive isotopes want to relocate.

Visitors to the site gravitate, inevitably, to the spectral remains of Pripyat, a town of forty-nine thousand evacuated within thirty-six hours of the accident. It sits just five miles from the reactor, a Pompeii without

the people. Scattered belongings in empty apartments serve as traces of lives abandoned midstride. Radiation readings in the area have long since leveled off, but its residents will never return.

However, the city seems to have acquired a new half-life as a dystopian theme park of the late twentieth century.

Decades after the catastrophe, a scientific conference in Ukraine issues a declaration: *the complexity, importance and diversity of the long-standing problems that have arisen as a result of the accident make it necessary to support a high level of scientific research both now and in the future.*

It will have to be an international effort.

Scientists now know that *an accident anywhere is an accident everywhere.*

Abruptly, three astonishing science stories break on the same day.

A new development comes to light in a notorious bioterror case that's been dormant for years. A top bio-defense expert working in a government lab has killed himself, just as he was to be arrested and prosecuted for the crimes.

The case had been moving so slowly that many had forgotten about the five letters, laced with anthrax spores, that were sent through the mail to congressmen and media figures shortly after the attacks of 9/11.

Five people died and seventeen more were sickened, including postal workers who unwittingly handled the material on its way to someone else.

The news report brings it all back.

People in that tense fall of 2001 are afraid to open their own mail. They don't have special biohazard decontamination tanks at home.

Drugstores across the country sell out of Cipro, an antibiotic used to combat anthrax. No one who buys the drug cares that it blocks bacterial DNA replication by binding itself to the DNA gyrase enzyme, causing double-stranded breaks in the bacterial chromosome; they just want something that protects them.

The U.S. Postal Service announces guidelines for citizens without lab facilities or antibiotics. First off, open your eyes. Look for anything unusual, even in that pile of junk mail. Beware of any envelope with no return address, misspelled names, excessive postage, or traces of unknown powders or suspicious substances.

If you see something odd, *isolate it immediately.*

The FBI launches an investigation, code name Amerithrax. It will devote hundreds of thousands of agent-hours to the problem and interview nine thousand people.

But the Bureau can't work the case by itself, not beyond the basics. It needs the help of experts with advanced degrees in chemistry, biology, and nuclear chemistry.

One of those experts is the eventual *person of interest* himself.

DNA sequencing and radiocarbon dating of the anthrax traces in the victims establish that the cultures, no more than two years old, came from a strain first developed in a military research institute. The water used to create the spores came from the northeastern United States.

Six hundred mailboxes on the street in the target area are tested for anthrax, and one comes up positive. It's likely that something untoward was mailed from that location.

But the leads dry up, so much that the FBI posts an apologetic Fact Sheet on its website as the years pass.

While no arrests have been made, the dedicated investigators who have worked tirelessly on this case, day-in and day-out, continue to go the extra mile in pursuit of every lead.

But that may not be enough to placate an anxious citizenry. Something more has to be added to the ante.

While not well known to the public, the scientific advances gained from this investigation are unprecedented.

It takes some time but investigators eventually settle on a suspect, only to backtrack after it becomes clear he's been falsely accused.

After the latest suspect is publicly identified, upon his death, the case is still murky. Colleagues and friends believe he was incapable of such acts (it's not pertinent, from their perspective, that he held a patent on an anthrax vaccine, or that his own therapist considered him deeply troubled). No lab test can establish state of mind or motive, and everyone knows the FBI has been wrong before.

All one can say for certain is that no deadly letters have been mailed in several years.

And no one's giving up Cipro just yet.

In the same news cycle, researchers announce two other major developments.

Studies conducted on mice have shown that it may be feasible to develop a drug that mimics the effects of exercise. Scientists have produced genetically engineered mice that grow more muscle and have greater endurance than ordinary mice. Since the same gene controls muscle tone in both mice and humans, it may be possible to extend the findings (after more research) to people.

The public doesn't ask for additional information about the PPAR-delta protein; it just wants the pill.

And NASA scientists announce that deposits of water have definitively been found on the surface of Mars, with huge implications for the possibility of life on that distant world. They've decided to extend their experiment and will direct the Mars robot to search for chemicals containing carbon.

It's been a busy twenty-four hours. Science continues to accelerate on all fronts, inducing a kind of whiplash in even the most casual observer.

21

◎ Thought Experiment

Imagine, if you can, a world without science. In this alter-nate vision of reality, we never left Earth's surface. We've never contemplated ourselves from afar, or considered what it might be like to expand our limited perspectives on life. We

are anchored to the immediate, at the mercy of forces we make little attempt to understand.

There's no Encyclopædia Britannica, *but that's okay, because we're not especially curious either.*

We still have art, which some love, and a variety of beliefs, which many value, but the world around us is opaque, and our capacity for wonder has been severely diminished.

Some are comfortable in this constricted scenario, because genuine wonder invites responsibility. There's some work involved, and that's not always welcome. And sometimes understanding only adds to one's unease.

Others find themselves drawn to mysteries, invigorated by the very act of asking questions, but in a culture of indifference such an attitude is difficult to sustain. At a certain point it's easier just to let the questions go.

Question: Assuming all that, is this a world you'd want to live in?

Science tells us that every hundred thousand years or so an asteroid or errant comet crashes into Earth, and it's only a matter of time until an epic collision occurs once again. Astronomers are keeping special watch on the asteroid known as 1950 DA, which on its current path is expected to hit Earth in March 2880.

Science also tells us that our oceans will begin to boil in just two billion years, as our sun flames out and scorches Earth. At that point, if it hasn't happened earlier, life will be extinguished, whatever life in the future looks like.

So it seems our days are numbered.

But in the meantime, there is much we haven't yet discovered, and much we just don't know. It's tempting to leave such efforts to experts. But the rest of us ought to do our part as well—we ought to learn their language, try to make contact, if only to understand what's at stake as the defining enterprise of our age unfolds.

It could be pleasurable, and might even be prudent. Because the future, like it or not, is coming right at us.

Acknowledgments

First I must thank friends and colleagues whose early encouragement helped sustain me, particularly Cecilia Rubino, Mark Statman, Alan McGowan, Elaine Savory, Abigail Franklin, and my sister, Diane Brooks. I also wish to thank my publisher, Jack Shoemaker, my editor, Roxanna Aliaga, Laura Mazer, and everyone at Counterpoint for their skill and discernment. This book is stronger because of their efforts, and it was a joy to work with them. Thanks to The Corporation of Yaddo for the Iphegene Ochs Sulzberger Residency, and Eugene Lang College at the New School for faculty research and development funds. Thanks also to Trevor Harris, Stanley Kauffmann, and my agent, Georges Borchardt.

Finally, I would like to dedicate this book to the memory of my beloved nephew, Travis Mulligan. He would have been its most avid reader.

Selected Photo Credits

Author Bio

COLETTE BROOKS's essays and creative nonfiction have appeared in several publications, among them *Partisan Review*, *The New Republic*, *The New York Times*, *The Georgia Review*, and *Hotel Amerika*. Her book *In the City: Random Acts of Awareness* won the PEN/Jerard Fund Award in 2001. She has taught at Harvard University and the New School, where she is currently a faculty member. She lives in Brooklyn.

Printed in the United States
by Baker & Taylor Publisher Services